Jillaroo

Jillaroo
STATION LIFE IN THE OUTBACK

SUSAN COTTAM

VIKING O'NEIL

Viking O'Neil
Penguin Books Australia Ltd
487 Maroondah Highway, P.O. Box 257
Ringwood, Victoria 3134, Australia
Penguin Books Ltd
Harmondsworth, Middlesex, England
Viking Penguin Inc.
40 West 23rd Street, New York, N.Y. 10010, U.S.A.
Penguin Books Canada Limited
2801 John Street, Markham, Ontario, Canada L3R 1B4
Penguin Books (N.Z.) Ltd
182–190 Wairau Road, Auckland 10, New Zealand

First published by Penguin Books Australia Ltd 1990

10 9 8 7 6 5 4 3 2 1

Produced by Viking O'Neil
56 Claremont Street, South Yarra, Victoria 3141, Australia
A division of Penguin Books Australia Ltd

Designed by Sandra Nobes
Cover design by Ron Hampton
Typeset in Palatino Roman (Linotron)
Printed in Hong Kong through Bookbuilders Ltd

National Library of Australia
Cataloguing-in-Publication data

Cottam, Susan.
  Jillaroo: station life in outback Australia.

  ISBN 0 670 90254 3.

  1. Cottam, Susan – Correspondence. 2. Jackaroos –
Queensland – Correspondence. 3. Ranch life – Queensland.
I. Title.

636.201092

Contents

References to weights and measures are given in the way they were originally expressed, in imperial, to retain the true flavour of place and time in which these letters were written, even though the change-over to a metric system occurred during this period, in 1966. Australian currency, however, has been expressed in the metric system.

inch	= 25.4 millimetres
foot	= 30.5 centimetres
yard	= 0.914 metres
mile	= 1.61 kilometres
acre	= 0.405 hectares
square mile	= 2.59 square kilometres
°F	= $\frac{9}{5}$°C + 32
mile per hour	= 1.61 kilometres per hour
pint	= 568 millilitres
gallon	= 4.55 litres
pound	= 454 grams
stone	= 6.35 kilograms
point (rainfall)	= 0.0254 millilitres
hand (horse height)	= 10 centimetres approx.

Introduction

In August 1965, while tapping away at the latest in electric typewriters in the Elgin office of a well-to-do architect, I stared gloomily out of the big window at the dank, dripping trees wildly tossing their water-logged branches in the half gale that was blowing. A couple of bedraggled collared doves were hunched in the doubtful shelter of some glistening rhododendrons, looking as depressed as I felt. This was summer?

'I'm going to emigrate', I announced to my fellow secretary, who was doing the wages for the nine members of staff. She looked up, said absently, 'Oh yes?' and resumed counting out pound notes.

When one o'clock came I put on my green mac, tied a scarf over my head and scurried up the cobbled street to the tiny travel agency on the corner, where I gathered all the available leaflets and information about Australia. Then I returned lunchless to the stuffy warmth of the office and settled down to browse.

The outcome of that soggy summer's day was that I filled in various papers and sent them off to Australia House in the Strand, electing to go to Queensland to start with, simply because I'd learned that most migrants went to New South Wales and never moved out of Sydney from then on. I also wanted to work in the hot, dusty environment of a cattle station – though don't ask me why. Probably the result of watching too many westerns on television; the prospect of riding for miles across unfenced country must have fired my imagination.

It seemed ridiculous, being able to travel from the north of Scotland to Brisbane airport for the princely sum of £10, but I took full advantage of the offer.

It was February 1966 when I boarded a BOAC 707 at Heathrow, clutching my dog-eared Junior Jet Club Log-book, a remnant from earlier days when I used to fly from Ceylon (Sri Lanka) to boarding school in Scotland and out again for the long summer holidays.

This trip, however, was the most adventurous. I didn't know a soul in Queensland, had no job to go to, and had roughly £40 ($80) in my handbag. An old friend of the family in Ceylon, Dorothy Gordon – herself an expatriate Aussie – had written to tell us that her good friend Mrs Sinclair would meet me at the airport and put me up for a few nights while I looked around.

This book is made up of extracts from letters written to my parents in Scotland. I shall apologise once only for the naivete that crops up here and there, but then (to quote a popular song), I was only nineteen.

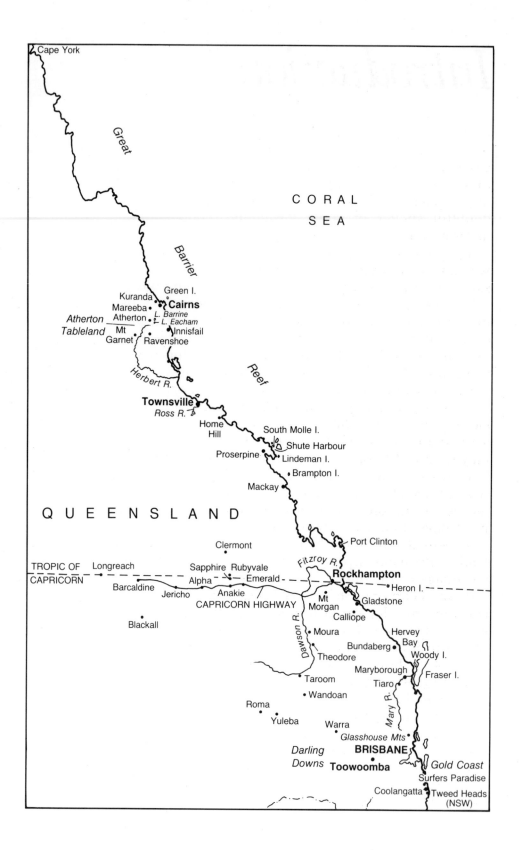

Part 1

Wavell Heights, Brisbane

Thursday, 3 March 1966

Well, here we are then! The plane arrived exactly on time at 7.05 yesterday morning, having lost fifteen minutes on the way from Manila, and we flew through an alarming storm over the Gulf of Carpentaria with lightning darting all over the place. I slept about three hours altogether on and off – mostly off, because some wretched infant cried all night long. I decided to miss the rush for the bathroom, so went along at 4.30 and had a long chat with a nice steward who told me he has a friend with a huge cattle station (80 000 acres) near Yuleba, and that I must go there if I'm in need of a job.

The customs man wanted to look at my riding boots in case they had been on any farms with foot-and-mouth, and then an immigration officer introduced me to Mrs Sinclair, who was waiting patiently, and we came out here by taxi. She is a very kind, pleasant lady and put me to bed right away because my head was spinning and my ears were sore. I slept until midday and woke up soaked in sweat and wondering where on earth I was. However, a large glass of iced pineapple juice on the bedside table saved my life.

In the afternoon Mrs Sinclair's married daughter came to take me swimming in a pool near by, which was lovely; I felt very conspicuous, though, with my legs leek-white from wearing woolly stockings. Fiona, the daughter, has just moved house from Darwin with her husband and three boys. The oldest boy is about nine, and has a wonderful accent, *I* think. However, it seems that was the main reason for their move!

Last night was comparatively cool after the heat of the day, and I slept like the dead with one blanket under a mosquito net – shades of Ceylon. Still feel a bit tired, mostly because we took a tram into the city today and went to the immigration office. There I talked to a chap who said I could probably get a good job as a governess or family companion, but that a 'jillaroo' would be more difficult. Because of the drought a lot of men are out of work, and they obviously get preference over girls.

We then trailed round some of the stock and station agents, but they all said much the same, though Elder Smiths thought there was some hope of a jillaroo's position, probably nearer civilisation than way out west. Mrs Sinclair's friend Mrs Golden has a cattle property at Yuleba, only 300 miles away from Brisbane, and she has invited me out to stay for a week or so to see how things are done. Maybe I'll be able to find a job from there.

You would love all the tropical flowers and trees that grow here – rampage might be a better word – frangipanni, tulip trees, flamboyantes, jacarandas, mangoes, plus all the citrus trees, of course, and countless other beautiful things. The only wildlife I've seen so far, though, is one tiny lizard, a few sparrows and a large bird that flapped off a fence. It could have been a kookaburra but was more likely an owl. There's a 14-year-old dog in the house called Suzy, who is a little peculiar due to her age, but quite affectionate. She likes to amble round the block with me for some gentle exercise. The house is built to catch every breeze, which is a good thing, because yesterday the temperature climbed to 86°F, and the humidity was almost as high. It fell to about 60° at night, but autumn is approaching, so it should cool down quite a bit.

Mrs Golden rang from her place, Potter's Flat. Mrs Sinclair was in her bath, so I answered and received a lovely welcome to 'this sunburned land'. She sounded very English compared to everyone else out here. Anyway, I've bought a ticket for Yuleba dated for Monday next at 8 a.m., which cost 6 dollars 50 cents.

Potter's Flat, Yuleba

Tuesday, 8 March

I met a cow this morning whose tail had been bitten right off by a dingo. You will gather that I am in the bush at last and loving every minute of it, even though this is only the first day. This morning I was up at 6.30 (Mrs Golden rises at 5 a.m. every day and makes a pot of tea) and went out to watch her son Hal doing the milking. He gallantly allowed me to try, and I succeeded with the nearest two teats, but the further away ones kept missing the bucket and hitting the dust. Also I was so slow that the cow quickly 'hung up' on me and refused to give any more milk.

After breakfast at seven of fried steak and gravy, a bunch of grapes and a glass of creamy cold milk, I was introduced to Archie, the silent stockman, and he, Hal and I went out on horseback to muster a paddock of 3000 acres (some paddock!) to look for any cattle that hadn't been 'bangtailed' – in other words, those that had missed the last muster, when the tufts at the end of their tails are removed with a sharp knife – a simple method of keeping tally. I was riding a huge, speckled grey gelding who was very quiet and obedient (Mrs Golden's sub-normal daughter, Marianne, rides him sometimes). The saddle was really comfortable, with thigh pads to keep you on board during an emergency; they call them 'poleys' out here. We ambled along at a walk most of the time. The ground is littered with fallen ring-barked trees and spiked dead branches; in fact, the country could become boring, as it is entirely made up of gum trees, mostly dead, and hard yellow grass like standing hay – no hills or rivers, just a few dry creeks and eroded washaways. The cattle drink from dams or troughs supplied by windmills and bores.

Hal Golden is very knowledgeable about the local fauna; unfortunately it's rather difficult to hear what he says as he talks almost without moving his lips, and usually with his back to you. He met me at Yuleba station last night in a Holden utility. It took the best part of two hours to do the 39 miles to the homestead along a very interesting bush track, over creek beds, round trees and between solid anthills. At one time a wallaby hopped across the road, in the headlights, which of course delighted me. Hal says the only things to beware of in the bush are red-back spiders and the wrong kind of man. I have been warned! He and his attractive wife Mary are coming to the big house for dinner tonight. They have three small children and live in a cottage beyond the stockyards, while Mrs Golden and Marianne live together in the house.

I am burnt already by the sun – mostly during a weekend visit to the Gold Coast where I christened the Pacific Ocean. The swimming was heavenly, but with the ocean's powerful undertow and enormous breakers it was also a trifle alarming. I went with Mrs Sinclair's daughter Fiona, her husband Peter and three children, and had lunch with friends of theirs in a rented flat on the beach. The friends have a sheep property some 600 miles north-west of Brisbane, near Barcaldine, and have asked me up to stay for a month if I am at a loss for somewhere to go.

There's a funny little Australian terrier here at Potter's Flat who adopted the Goldens a few years back and has been here ever since. She's very sweet, but will insist on bringing in dead snakes and bits of stinking meat that she buried months ago. Yesterday Mrs Golden shot a brown snake, which crept under the house to die. Today it was beginning to get rather high, so Hal removed it with a lot of hindrance from the terrier. Apparently there are lots of snakes around here, all poisonous; also several varieties of scorpion, centipedes and numerous spiders, including the aforementioned red-back whose bite has been known to kill the occasional toddler. I saw lots of wild cockatoos today, waddling about on the ground picking up seeds, also some lousy jacks, a darker equivalent of the Ceylon common babblers; a most unpopular bird in cattle country for their habit of loudly announcing the approach of men on horseback to unsuspecting cows, which in turn involves unnecessary galloping to retrieve the runaway beasts.

Marianne, the daughter, is 33, with a mental age of about twelve. One really can't help feeling sorry for her. She's not attractive in any way at all; she looks as though she's about to burst into tears all the time. At lunch today she said suddenly: 'Oh Mummy, I'm so glad Susan's come to stay with us. I won't be so lonely now'. I didn't know what to say. I don't think I shall be much good to her as company. She wears the dowdiest clothes – skirts and dresses almost down to her ankles – and seems to eat nothing but boiled eggs mashed up in a cup with half a loaf of dry bread. She's listening to Children's Hour on the ABC just now.

The train stopped at thirty-seven different places on the way out yesterday – no wonder it took so long. I saw a few mirages, and what I thought might be a wombat snuffling in the grass, but apparently they are very rare so it probably wasn't.

I must go and unpack now. Tomorrow I'm to help bring in a bull that needs its tail washed, of all things! One of my jobs will be to fetch the horses in the mornings, riding a little pony called Dolly, known as the night-horse (should be night-mare by the sound of it), who's kept near the house specifically to run in the other fourteen lovely animals. I saw them come down to water today, a beautiful sight with their glossy coats (never groomed) and long, slender legs.

15 March

I received your letter today with great joy, as you can imagine. How lovely it is to have news from home. Archie went into town to get the batteries for the lighting plant, so he picked up the mailbag, which usually comes tomorrow. So sorry to hear you still have that wretched flu – I just hope the doctor can do something for you. Thank you for the loan of your wristwatch, Mum. It keeps good time, and the strap is not too loose because the heat has swollen my wrists a little. I am nicely tanned already, though my arms and shoulders look like an old lizard having a hard time shedding its skin – the result of swimming at Surfers Paradise on Sunday. My hands were burned holding the reins the first day out mustering, but they're nice and brown now.

Last Wednesday I went out on Blue Boy – another grey – with Archie, as Hal was away in Roma seeing his cattle being slaughtered. We hadn't gone far when we came across a young cow lying down alone. She staggered up but fell again, so Archie went back to the house for a rifle, leaving me by the poor heifer, who hadn't had any water for about two days. She looked awful. When Archie returned I held his horse while he shot the cow between the eyes, and she went down with a grunt. Now I know what to do should the occasion ever arise again – God forbid! She had been damaged by having her first calf pulled from her, as she couldn't expel it by herself, a condition called 'obturator paralysis'.

We rode on to the Four Mile Paddock to look for a bull; we found four, but none was the right one. It's pretty wearisome riding because none of the horses are shod, and one has to walk most of the way to conserve their energy, weaving in and out of gum trees and stepping over dead branches. I have drunk more tea – and enjoyed it – since being here than ever before. It's so refreshing, especially without milk. Anyway, we returned to the house for lunch and Archie went out alone again later and found the missing bull.

My jobs about the place include feeding Dolly the night-horse, separating the milk each morning, washing the separator and reassembling it ready for the next day, a little bit of cooking and some washing up. At present I'm weeding the tennis court, a hot job that I only tackle before breakfast and after tea. It's full of couch grass and pigweed. Hundreds of tiny double-bar finches line up on the fence to watch, all pressed together and shoving each other further along the wire, and today there was a sacred kingfisher close by, diving every so often into the long, dry grass and emerging with a grasshopper. Makes a change from fish I suppose, and since there are very few permanent waterholes containing fish out here, he must have adapted to living off the land. Two rosella parrots sat on a tree close to my room this morning: beautifully coloured, with sky-blue chests, bright red vents, dark grey wings and lime green and black on their backs.

No one works during the heat of the day. We get up at six every morning – it's almost chilly then – and I go out to watch Hal do the milking; that will doubtless be one of my many other jobs before long. Yesterday he made me bring in two of the cows and trap their heads between two posts of wood (a bail), then rope the hind leg nearest to the person milking, tying it to the fence to prevent the cow kicking out. The calves were tethered near by – they are kept in all night to let the cows make enough milk for the homestead – and I had to untie each cow's calf in turn, let him suck just long enough to get the milk flowing and then tie him up again.

Poor Marianne has dreams about her father, who died some time ago, and most days she corners me to tell me all about them. It's rather awkward, as I never know what to say in reply. She spends hours in the smallest room (which is, by the way, a well-built shed in the garden with a wooden seat over a seemingly bottomless pit, rather draughty and frequented by the most repulsive bluebottles) and she knows every single telephone number that her mother uses. No one needs the directory while Marianne is around.

Mrs Golden is a lively seventy-one. It's quite wonderful how she works away virtually all by herself – but then she loves the life and is fiercely independent, so doesn't want to be helped in any way. All I can do inside the house is wash up occasionally. Incidentally, a little jade-green frog startled me somewhat the other day by coming down the *hot* tap straight into the sink. He was still very much alive, so I carefully put him out among the pot-plants.

On Friday I had a great day learning how to draft steers from heifers, cows and calves and put them all into different pens ready to go through the tick dip, which they did after the stock inspector had looked them over. He was a young fellow, and called me 'Sue' right away. At the end of the day Hal said I had done quite well in the yards, but he warned me to take more care when I'm behind the horses. They say one learns by one's mistakes! He's a patient man, full of droll humour and pleasant jokes; his definition of an English saddle is a 'self-emptier', and he reckons that they'd be useless for a hard day's mustering. The only disadvantage I have found to the Queensland 'poley' saddle is that owing to the thigh pads, one cannot stand up and stretch too well. The knee I twisted two years ago gets a little sore if we're out all day.

After lunch the cattle were loaded onto a huge float, but just when they were all safely aboard the driver discovered a puncture on an inner back wheel. I departed before the air became too blue, and Hal came back half an hour later covered in liquid manure, as the bottom of the float is slatted to let the air in and the dung out. We all fell about laughing, which didn't exactly please the sweet-smelling apparition.

I have just been down to the hen run with Mrs Golden, where she shot a sand goanna that was peering smugly out of a hen's nest, having eaten all today's egg supply. There were two others up near the cattle yard, climbing a tree, and today a grey kangaroo came hopping along quite close to the house. I had visions of herds of 'roos, but it seems they go about in small family groups and don't appear very often during the hot hours of the day.

At night after supper the three of us play cards or read, falling into bed no later than nine o'clock. Last night was stifling, with no wind at all. I woke soaked with sweat to the noise of a pack of dingoes howling near by: a spine-tingling sound, but rather pleasant in a weird sort of way. They are terrible pests though, doing a lot of damage to calves and sometimes older beasts, and are shot or poisoned without mercy.

Meals here consist of steak for breakfast, cold meat and salad for lunch, and a hot meal – stew, curry or a roast – at about seven in the evening, with masses of vegetables. The only things that manage to grow in the straggly garden are lots of capsicums, quite a few cucumbers and things called 'chokos', which are a bit tasteless but good if drowned in melted butter. In winter Mrs Golden grows daffodils, of all things!

The Goldens are very non-class-minded and have no racial prejudice at all. In fact Hal said he would rather have his children marry Negroes than Roman Catholics. They had some African veterinary students to stay last year, and seemed to have grown very fond of them. Ian Smith, they think, is an awful man. It makes a change to meet such different views from those we hear in Britain.

There is no sign of rain, and the ground is dry and cracked. Of course the wet season has passed and there wasn't much rain anyway this year, but I have had three baths and washed my hair twice with water pumped up by the windmill.

Mrs Golden has been very good about ringing up people to enquire after jobs. One of the replies was, 'My wife won't let me employ jillaroos'. She wants me to go to people she knows and likes to begin with so that I can get the feel of things before moving on to something more permanent, and suggests a horse or cattle stud rather than a station, as there would be less maintenance work to do and more riding – perhaps somewhere on the Darling Downs, which is beautiful rolling country.

Poor Dolly is nosing around very despondently at the dried-out grass roots in the home paddock, so I must go and cut her some decent fodder.

22 March

I am simply dead beat tonight, so you will have to excuse the scrawl. We've been out all day mustering – my first complete day in the saddle. We left the yards at 9 a.m. to collect 109 head, which were brought back and yarded; then we had a quick cold lunch and went out to find a further 122 animals to bring back. We had tea at the yards beside the milling shorthorns and Herefords, then let them all out again through a different gate to go down to the dam and drink their fill; they'll stay in that small holding paddock for the night. Tomorrow we shall round them all up again for dipping, but first have to find another fifty or so absentees, which are no doubt hiding in the thick belts of scrub in the Four Mile.

My bottom is rather tender – Hal says I shall now appreciate why that cowboy series on television was called 'Rawhide' – and my hands are burnt again. Blue Boy certainly came to life when the cattle got into the scrub. It was wonderful, he knew just what to do, and was so good and patient – I really learned a lot from him. Hal told me to go and head off the leaders of the herd who were running in the wrong direction, so we had to gallop to catch up with them. How that horse managed to avoid falling or even stumbling, I'll never know. Four fallen trees lay parallel, hidden in the grass. There was no room for his feet to go between, and no time to swerve, so we soared over them – both together, thank goodness – and successfully turned the leading bullocks. Later I was ordered to go and fetch a small mob of cattle and keep them together on the fence. Of course some decided to stop and stare while the rest headed for the shade trees on the creek bank. Fortunately help came before I lost the lot.

Last Wednesday there was quite a storm, but not much rain. However, on Sunday night I was awoken by a clap of thunder that made the house shake. The thunder went on cracking, there was non-stop lightning, both forked and sheet, which lit up the whole place, and the wind was blowing a gale. Suddenly rain sluiced down as if someone had upended a gigantic bucket; 94 points fell in half an hour, and the next-door property had 600 points, which caused a lot of damage. Still, everyone is thankful to see the water. The grass will shoot now.

I made a fool of myself milking on Thursday. After a lot of hard work on my part I'd managed to accumulate about two inches of milk at the bottom of the bucket when the old cow let fly with her free leg and kicked the pail from between my knees. I lost the lot, of course, but Hal was very kind and told me not to worry. In fact he says I'm getting quicker and better at milking all the time, which is encouraging. The Herefords are fairly easy to milk, actually, having good big udders, but the shorthorns are so small it's difficult to get a proper hold.

I've been doing a lot of weeding and raking up cut grass this week, and have graduated to working the hand-driven separator alone, now that I know at what speed it has to be kept.

There's a big, noisy fellow called Dick Perrett staying at the moment. He has a cattle spread a few miles away, but has come over to help Hal for a couple of days. He's a great character, full of interesting tales about the bush. I drove his Holden all of a quarter of a mile to Hal's house, managing quite well despite there being only three gears on the column.

27 March

Once again I'm worn out and thinking of my bed with great longing – and it's only eight o'clock.

We went out to lunch with Dick Perrett, who stayed here last week (he would insist on calling me Judy, for some unknown reason). He has a lovely bungalow and a superbly run property, which he took us round to see after lunch. We all piled into his Holden, which he treats like a Landrover; it even has oats growing under the dashboard and cobwebs in the back windows.

Hal, Archie and I are going back to his place, Bottle Tree Hills, tomorrow to help muster for about three days, and I am to spend the nights with Dick's brother and family who live 'next door', 3 miles away. He says I can borrow the Holden to get myself there and back – I'm to be here for breakfast in the morning. There's only one rough washaway to negotiate, so I should manage, and it will give me experience of driving on the bush roads. Dick doesn't think I'd care to stay the night with three men, however nice they may be.

I have been promoted to riding Dolly out at 5 p.m. each evening to bring in the milkers, and am very pleased, because the bush is at its best at that time – all green and gold and silver, with long blue shadows – and the sun has lost a lot of its heat. Dolly may be small but she's as skittish as a 2-year-old and can outwalk most bigger horses. Incidentally, one of the horses was missing last week; Hal eventually found him dead under a bank that had toppled over after the last big storm. He thinks he was struck by lightning.

Last Wednesday we went mustering to find fifty head, but only collected forty. A violent storm broke over us on the way home, and we were all soon soaked to the skin. Boots overflowed, water poured down our necks, the horses skidded about, the cattle became restless – it was like living a scene from a western movie, but without the cameras. We were to have gone out in the afternoon to find the missing ten cattle, but the sky clouded over, turning an ominous green and black, and soon let its load down on Hal, Archie and me as we huddled in the saddle shed trying to keep warm. The rain turned to hail, great chunks of ice nearly an inch in diameter being hurled down on the corrugated iron of the shed – we couldn't hear ourselves think. Nearly two inches of rain had fallen by the end of the day, and now the bush is lovely and green, the grass really quite lush in parts.

Mrs Golden is going to pay me $3 a day for weeding the court. I was quite amazed, but she thinks it's only fair.

On Friday I went out mustering again. Dick was with us, and when he dismounted to retrieve his hat, which had blown off, his horse bolted. More by luck than good management I happened to get alongside the runaway, leaned over and caught his bridle, for which Dick was most grateful; otherwise he'd have had to walk home. Before lunch we brought in a big herd of some 300 cows, plus all their calves, and yarded them for dipping. Poor things, they do hate being man-handled into the deep concrete bath full of stinking tick-killing insecticide. One old cow dug her heels in and refused, then went quite mad and charged everyone she saw, frothing and slavering, her eyes red and wild. She had to submit eventually, though, and was pushed through the dip to stand shivering in the draining yard, where all the excess dip mixture runs off and back into the tank.

At one stage two bulls started fighting close to me, lumbering around and bellowing, and banging their great curly polls together. Hal shouted at me to leap for the rails, but I was already halfway up. Mrs Golden brought us tea at four, complete with rock cakes and gingerbread, which was very welcome, and the dipping was completed by suppertime.

The four of us were mounted early next morning, ready to let the herd out. We drove them through the trees for about half a mile until we came to a wide plain, where they were halted, and the sorting out process began. We had to match each calf with its mother, because they had all become separated during the chaos in the yards. It was a terrible job; the poor little beasts lagged behind, bawling hoarsely. Their instinct told them to go back to the place where they last had a drink of milk – mum should be there – so of course they were trying to double back to the yards all the time. I found myself chasing one little chap who made a break past me, hightailing it for the yards. Every time I tried to come abreast of him he went faster, bolting through clumps of brigalow saplings, gum trees and huge, thick spider-webs that plastered themselves to my face and the horse's nose and ears.

The sweating trio was nearly at the yards when the horse finally caught and turned the calf, who shambled back the way he had come, complaining pitifully, his tongue hanging out. Back at the herd a few anxious cows came forward to look him over, and luckily one claimed him as her own. He was too exhausted to suck right away, but he soon stopped panting and nudged his mother's full bag with great enthusiasm. Most of the cows had found their offspring, but it was well into the afternoon before we were satisfied that all were paired off.

Hal says he must doctor the dead horse with strychnine, to poison the dingoes. He explained that there are so many dingoes about, and they take such a terrible toll of the calves, that they have to be exterminated by whatever method is available. I see his point, but basically I still can't accept it. An awful lot of other animals must suffer by picking up the poisoned baits, especially crows, cats and wild pigs. (But then these animals do a lot of harm too.) Apparently some stations drop baits from low-flying aircraft all over the country, and that to me seems quite monstrous.

5 April

To answer your questions:

1 The house at Potter's Flat is a large, square, blue-painted wooden bungalow set on wooden blocks to deter white ants and allow the air to circulate underneath. Inside there is a central dining room with a door at each end opening onto steps that lead down to the garden. There are two bedrooms off one side, one bedroom and the kitchen off the other, and a small bathroom beside my room; the latter is small but light and airy, and contains two beds, a dressing table, wardrobe and chair. The window faces west, looking out into the endless eucalypts.

2 The tennis court consists of the hard red soil rolled and beaten level with a layer of sand on top. At present, too, there's a large goanna's hole near the middle. Hal and Mary sometimes play, but they haven't used the court for a while, hence the weeds and goanna diggings.

3 The fresh vegetables come every week with the mail truck, which also brings drums of oil and diesel, fencing wire, bread and other groceries. Mail day is quite an event in these parts.

4 No, I haven't seen the Flying Doctor.

Last Monday Hal drove me over to Dick Perrett's place to muster; Archie came along later. I was given an ancient bay gelding to ride. He was all of 25 years old, with feet like soup plates over which he stumbled most of the time. He was exchanged for a keen young grey in the afternoon, much to my relief.

The mustered cattle were yarded before noon, when I made lunch and pints of tea for the men on the wood stove in Dick's bachelor kitchen. We ate out on the cool verandah, sitting in comfortable squatter's chairs and listening to the baleful bellowing coming from the yards.

After a short rest it was back to work, drafting the calves from their mothers and pushing them into the crush with the help of an electric prodder. Archie was in charge of the 'cradle' at the top end of the crush, ready to catch each calf as it came out; he would trap it deftly in this machine (which looks like an outsize waffle-maker) and pull it over to lie on its side, there to be branded, earmarked, castrated if male, and vaccinated against blackleg. All this happened in the space of about a minute per calf; then the cradle was opened and the poor little beast could jump up and totter away across the yard. They looked really pathetic, standing there dismally recovering from the trauma, some of them with blood trickling down their back legs. My work consisted of keeping the crush full of calves, making sure the syringes were full of vaccine and washing the castrating knife in a tin of disinfectant between each calf.

Later on in the afternoon the last calf was released. We had a quick and very necessary cup of tea (not exactly Rockingham china; we drank out of chipped enamel pannikins, made in China) before climbing on our horses and escorting the cows and sore calves back to their paddock some miles away.

That night I spent at the home of Ted Perrett, a brother of Dick, where I shared the bed with a cat and a fat green frog. I had to borrow a nightie from the lady of the house, Jean Perrett, because all my clothes were at Kabunga West, the property belonging to Owen Perrett, the first brother. This was where I had planned to stay originally. It also meant dressing next morning in the same filthy clothes I'd worn in the yards the day before. Dick left his car for me, and had a lift back to Bottle Tree Hills with Hal, so I set off at sun-up in some trepidation. I managed reasonably well, in fact, though I was glad there was no one watching but the magpies and butcher birds. I arrived just as the men were catching their horses. I was given a huge black horse called Tanto and we split into pairs, Dick and I going in one direction to collect some forty head of weaners, while Hal and Archie rode off elsewhere.

There was more branding and vaccination to do that afternoon. I plucked up courage and inoculated four calves myself, but I didn't enjoy it at all – I'd make a rotten nurse. Then at sundown we all set off at a canter to muster the bullock paddock before dark. Hal muttered darkly something about a 'Yankee round-up' – normally Australians like to muster very slowly to save running the fat off the cattle – but I must say it was great fun going at speed for a change. The bullocks didn't know what had hit them, but they obliged by stampeding down the slope, with us in hot pursuit, and yarded themselves for the night.

I washed in three inches of cold water in Dick's bath, with the help of an oil lamp and sundry beetles and moths. We then drove over to Kabunga West, where Owen Perrett and wife Jessie live. They are such nice people, kind and friendly, with two super sons, Rodney (19) and Bruce (16), and a daughter Ursula (14). Rodney took me straight out to the back where he keeps his car, an old, bright red Ford Anglia convertible he's very proud of, and showed me the odd assortment of vehicles they have. Among them was a Model A Ford in working condition and a Ford wreck which Bruce drives. They seem to be just as car-crazy out here in the bush as back in Britain. Rodney also asked me to go to a dance with him and some friends on Saturday night.

There are four black and white cats about the place – Burnt-cake, Biddy, Tommy Tommy and Pin-up – definite Aboriginal influence there, I should say. Also a parrot and two cattle dogs. Talking of Aborigines, there are five working here, as well as Coral Sea Islanders; they work reasonably well for a week or so, drink their money up in metho and rum, stay drunk for a fortnight and then drift back to work to make more money to buy more grog . . . and so it goes on, a tragic circle with no happy ending. I'm going to help cook for the men next week and pack their saddlebags each morning.

Next day I drove to work again. Big Tanto helped me muster a 2500-acre paddock, and took off after three steers that beat him to the scrub, where they separated in different directions. Luckily Hal spotted my dilemma and came to the rescue. After lunch the steers were drafted off and injected with blood against tick fever; I was kept busy reloading the syringes again. Their horns were tipped by a ferocious cutter, and blood spurted all over the place in jets from the severed arteries. As I was allotted the task of releasing the bleeding animals (no, Mum, I'm not swearing!) from the crush, I became covered in blood; it went all over my face, hat, shirt, boots and jeans. The cattle must lose a fair amount before it eventually coagulates.

Mrs Perrett was at the gate when I drove up in the evening. She took me by the hand and led me to the bathroom, where she'd prepared a hot bath complete with a violet-scented bath cube. (I didn't think it was so obvious!) Biddy and Burnt-cake stayed most of the night with me on the bed beside a big open window and watched with cool disdain while I counted my wounds; fifty-six mosquito bites, twenty-seven bruises of varying shades and one septic blister on a finger, caused by continuous tugging on the reins for three days. I don't suppose it helped much getting poison into it while the dip was being filled.

Next day Dick lent me his best and favourite mare to ride, the ultimate compliment apparently! She was a delight to work with: keen and alert and a very fast, smooth walker. The mustering finished by midday, I returned to Potter's Flat with Archie, while

Hal followed in the Landrover. There I washed my hair, packed my possessions and after a cold lunch was driven along the road by Hal to meet Mrs Owen Perrett, who was going to Miles to collect Bruce, her younger son, from college. It was a two-hour drive, mostly along sealed roads for a change.

On Saturday morning Rodney took me into Wandoan in his old Anglia. The town is some thirty miles from the homestead, and after the first two miles we were both covered in dust as the car is full of holes and rattles along at a top speed of 40 miles per hour. Having hit town, we walked dustily into the cafe, devoured a super sticky fruit sundae each and met some of Rodney's mates. It didn't take them long to convince him that he was in great need of a beer, but as the drinking age limit here is 21, we had to drive well away from the town. Two car loads of us left the road and 'went bush' where we drank cool, refreshing, bottled beer that had an unexpected kick to it. Quite a little party developed, and by the time we trundled home again there was only an hour left in which to get ready for the dance; it was being held at Warra, 120 miles away, and we had to leave at 4.30 p.m. In that hour I had lunch, washed my hair and changed into that blue dress with white flowers round the neck, *stockings* and white high-heeled sandals.

We set off in Mrs Perrett's new Holden, with Bruce driving, as Rodney was feeling drowsy after his midday beer session. Although he's only 16, Bruce is a good driver who doesn't show off like a lot of young fellows do. Mind you, he's been driving since he was knee-high to a grasshopper, so has had lots of practice. We stopped *en route* to have tea at some friends' house and pick up their son Peter to take him with us, so it was 9.30 by the time we parked outside Warra's brightly lit hall.

Inside there was chalk on the floor, the boys were all dressed in white shirts with the sleeves rolled up – some wore ties and some didn't – and dark trousers, while the girls had on anything from sequinned long dresses to short cotton shifts. Anyway, everyone looked smart and tidy and out to enjoy themselves. Bruce and Peter were hoping to pick up some interesting girls for the evening, but had no success and had to dance with me. The country dances are a lot of fun as there is so much variety, not like the endless 'shake' sessions back home: gypsy tap, pride of Erin, Boston two-step, Oxford waltz and of course good old rock 'n roll. Rodney is a terrific jiver, so we worked up a lather happily until 2.30 a.m. when the band went home.

We got home with the dawn at 5.30. I went to bed, but poor Bruce had to do the milking first.

Kabunga West, Wandoan

Easter Day

These charming Perretts left an Easter egg outside the bedroom door for me to trip over this morning. They're so kind in every way, and I feel thoroughly spoiled. Bruce fills the bathtub with buckets of hot water from the wood stove for me each night, and yesterday his father brought his pride and joy into the garden to show her off to me: a beautiful dappled grey, part-Arab mare with dancing feet and a lovely neat head. She is in foal to an Arab, so the result should be good.

Young Bruce prevailed on me to make a canvas seat cover for his 'car' – the wrecked Ford – to which I agreed in a weak moment. Actually it's not bad; at least it hides all the emerging stuffing and stray springs fairly well. This was the car that was rolled by five drunken Aborigines while they were working for Dick, so he gave it to his nephew to tinker with. It has no glass at all, the doors are jammed shut except for the driver's, which has to be kept closed with a piece of string, the bonnet and boot are missing and there are only two forward gears – but the engine is perfect! Bruce puts a hat over the exhaust manifold to stop the smoke blowing back through the non-existent windscreen. I haven't attempted to drive it as I'm quite sure it's definitely a one-man car and would not obey a mere female for an instant. Rodney, meanwhile, has gone off to the coast to trade in his little red Anglia for a more suitable station wagon.

On Good Friday Dick came over and took Bruce and me fishing for the day on the Dawson River. We took with us an extraordinary flat-bottomed boat made out of two Holden bonnets welded together in the middle – surprisingly effective, but hard to sit in without cushions. The Dawson is a wide, deep, dangerous river, full of swirling muddy water all the year round, but Dick knew of a quiet backwater between enormous trees where we strung a net from bank to bank when we first arrived in the morning. It caught seven bony-bream, two saratoga (a type of barramundi exclusive to the Fitzroy River system) and several smaller fish. Dick cooked one of the barramundi wrapped in leaves over a stick fire, Aboriginal style, which we ate for lunch, followed by bread and honey and billy tea. Bruce and I went for a swim, but the water was cold and murky and about fifteen feet deep in the middle, so we soon came out, dried off and lazed about in the boat, trailing lines over the edge. We didn't catch anything because turtles kept pinching the bait. Dick stayed on the bank and cleaned the netted fish, catching six large eels with the guts in about ten minutes. We had to leave before sundown as there was no road, and we had to find a gap in the fence to get through before dark. It was a lovely day altogether. I'll bet there aren't many inland cattle properties that had fresh fish for their Good Friday dinner.

The Aborigines have left for their Easter holiday. No one knows if they will return to work, or land up in jail, or be found dead or just flaked out somewhere. The Perretts employ them to keep them from full-time boozing, but the law says they have to pay them with money, which of course they immediately spend on liquor. Doesn't seem to make much sense really.

19 April

I have a definite job at last! It's on a 96 000-acre cattle station called Wendouree, near the little town of Alpha, about 300 miles north of here. The work will include mustering when they do any, housework and gardening, but it's mainly to look after a little boy of 3 – that is, if I meet with approval. The grazier is one Mervyn Carruthers, a widower, who lives with his crippled mother on the homestead 40 miles from Alpha, and there is a married couple already working for him. My wages will be $16 per week.

Last Sunday Bruce took me for a run in his battered Ford. We went rather fast over a gully and the petrol tank fell off. I had to help him tie it on again with wire – the first time I've ever been under a car, in the middle of the road too – and for some reason we both got the giggles and could hardly see for laughing.

Next day I did a half-day's mustering for Dick Perrett again, taking a mob of weaners along the fence from the yards to put them into a new paddock by themselves. This involved a certain amount of fast horse-work to keep them all together and heading in one direction, as they kept trying to turn back to mum. That evening I went up to the other Perrett brother's place, Cattle Downs, to spend a week there. A delightful cat called Mippy Mippy, heavily pregnant, shared my bed, sleeping between the bedspread and the blankets and purring like a little steam-engine. In the early morning just before dawn, however, two dingoes came right up to the house and howled dismally under the window. Mippy shot out of bed and I awoke with every hair on end, thinking they must at least be under the bed, the moaning was so loud.

While at Cattle Downs I went out mustering with Ted Perrett and Mark, their jackaroo, riding a lazy horse called Socks. Halfway through the morning Mark lent me one of his spurs, which had the desired effect on Socks; he flew about at a fast trot most of the time, breaking into a canter if he so much as heard the spur jingling. We camp-drafted (cut out of the herd) a few heifers, which was difficult and most trying. There is nothing as stubborn and flighty as a half-wild heifer when she wants to be – typically female! We stopped for lunch under some trees, lighting a fire and boiling our quart pots full of dam water for tea: proper billy tea, complete with floating leaves, twigs and the occasional mosquito larva, drunk with lots of sugar and no milk. As we set off again it rained a little, but not enough to do any good.

Later I was left at the yards, miles from anywhere, with all the heifers, to await the arrival of the stock inspector, while the other two rode off northwards. The inspector duly arrived – a morose kind of bloke – and the dipping was completed uneventfully, after which I rode home alone. Luckily Socks knew the way better than his rider did!

Dick came up for dinner. He ran over a porcupine (echidna) *en route*, which punctured his tyre as it died, so he felt justified in skinning and roasting it in the Cattle Downs oven when he arrived. We had it for supper and it was very good, though rather rich. It tasted like duck crossed with wild pig.

Next day we all went into Wandoan to decorate the hall for the big dance on Saturday. Someone strung up a mosquito net over the centre of the dance floor – filled with balloons and covered with gold and silver streamers – to which we added some finishing touches. Then we arranged bunches of eucalypt branches in suitable corners round the hall. For the big night itself I wore a full-length orange, woollen skirt, your white long-sleeved blouse, white shoes and my old Ceylonese moonstone scorpion. Everyone seemed agreeably surprised at the transformation from the usually dust-covered, be-jeaned, hard ridin' jillaroo they were used to. Dick was my escort and we had a lot of fun at the ball, but nice as he is – and I'm extremely fond of him – he's twice my age, so . . .

The bar was free all night, from 7 p.m. until 3 a.m.; Dick introduced me to Hal Golden's Uncle Fred, who was propping up the bar – a most amusing man. Supper was a buffet of prawn cocktails, curry, chicken stew, peaches and cream, and coffee – a super spread to which a lot of the people didn't do full justice, preferring their liquid diet of beer. Eventually our party returned to Cattle Downs at 4.30, where I set to and helped Jean get breakfast for eighteen people. She gave them barbecued chops, sausages and bacon, a trough of baked capsicum, tomato, sweet corn and onion and relays of toast. I got to bed thankfully at 7.30, slept until 11 a.m., had a light lunch and came back here to Kabunga West.

Rodney is back from the coast with his new car, and today he took me to their camp on the property where two of them are ploughing prior to sowing oats. He insisted on my driving the tractor once around the plot, ploughing at the same time. Well there's a first time for everything. I hopped in, Rodney showed me the gears and stayed aboard to see that nothing disastrous happened to his pre-cious machine, and away we went, ploughing merrily for about three miles, which was 'once around the plot'. By the time I'd finished I was covered in red dust, deaf as a post and all my teeth and bones felt as though I'd been picked up by some large animal and shaken violently. On the way back to the house we found a tree that had fallen across a fence. Rodney removed the tree, then taught me how to mend the top wire using a figure-of-eight knot. Your daughter is learning many things out here, some useful, some not, but all highly enjoyable.

Yesterday a neighbour turned up with a pretty chestnut horse with a badly lacerated neck, caused he thought by some loose barbed wire. Owen Perrett came out of the house with a bowl of disinfec-tant, cotton wool, and a needle and catgut, put a twitch on the horse's top lip to keep her still (they call it 'bushman's chloroform'), and sewed up the gash while I held the bowl. The poor beast didn't move a step, but stood trembling all over, her eyes nearly popping out of her head with fright. It seems most people in the bush treat their animals as far as they can without a vet; presumably vets are too expensive, and in any case would take too long to reach the station in an emergency.

Rodney has fixed up a long aerial for my little radio by stringing a wire from it through the open window up onto the roof, where he has tied it to a bent nail. Now she blares forth after a preliminary session of squeaks and whistles. The short waveband is most in use out here because it gives better reception.

Last Thursday the tractor, a nasty little machine with steel wheels and spikes, broke down at the camp near the oats, so Rodney took me with him in a big ex-army truck while his mate Colin (a half-caste) followed in a Landrover. When they had fixed the truck to the front of the tractor they left me to drive the 'rover, which of course wouldn't work: the clutch had gone. Rodney returned, put her in first gear and started her off, telling me to drive like that to the camp, which I did, very slowly without stalling. There he filled the hydraulic reservoir with castor oil from the first-aid box, which soon had the clutch 'running' smoothly again. I was then given the job of steering the on-tow tractor, which proved to be fairly simple until the chain snapped in a gully. The Landrover was put in front of the truck, attached by a second chain, in 4-wheel drive, the tractor was re-attached to the truck, everybody heaved and shouted at once and we flew up the far bank in a cloud of dust and fumes that enveloped *me* sitting in the tractor. Through sheer devilment Rodney and Colin drove fast over a raised cattlegrid, making the old tractor leap into the air and crash down onto its steel wheels again, which hurt more than my feelings but gave the boys much satisfaction.

Friday was the first day of the Wandoan show. Mrs Perrett took me in the Holden station wagon early in the morning before the crowds arrived, parking the car with its bonnet up to the ringside rails so that we had a good view of the camp-drafting event, which was in progress then. There was a bit of buckjumping later on, followed by the Feature Bull Ride, starring an evil-looking huge Brahman bull called Woorabinda Devil, which some unfortunate roughrider had drawn to ride. The 'Devil' dumped his rider before he was hardly out of the crush and bucked on round the ring alone before being ushered out.

Saturday dawned bright and sunny as usual. Rodney took me in to the show where we had a really good day, staying on for the ball after tea at the pub: curried sausage, roast chook and pumpkin, and ice-cream. Rodney bought a bottle of sparkling white wine – Australian of course – and we finished it between us amid fits of laughter. The blokes at the next table asked if we were just married or something, so we said that was more or less the idea, which struck us as exceedingly funny. Rodney is the first boy I've met with whom one can have a lot of fun without his becoming all sloppy and serious.

I changed into a dress in the Country Women's Association restroom, then we went up to the hall, where I danced every dance with at least fifty different partners. One very popular dance is the gypsy tap, in which the men continually change partners by tapping some other fellow on the shoulder and waltzing off with his girl – the same idea as our old-fashioned 'excuse-me' waltz. I met up with Terry Dredge, the stock inspector who came to Potter's Flat; also a good-looking chap who knew all about me although I had never heard of him. That of course is the trouble with being a newcomer.

By 2.30 a.m. Rodney and I were exhausted, so we left and called in at his Uncle Dick's place, where we had hot dogs, apples and masses of creamy milk. Dick was delighted to see us, even at that hour. We eventually got home at 4.30. I slept soundly until 9 a.m. and arose feeling remarkably fit, though rather stiff. Rodney didn't wake until 4.30 in the afternoon, when he demanded breakfast, thinking it was only early morning.

On Monday Dick arrived to take me to Theodore, from where I intended to catch the train to Rockhampton. It was a nice journey through the Dawson Valley, lovely ranges on either side, but the grass was brown and dead and the cattle were very poor, with staring coats. We reached Theodore at sundown and booked into the hotel. My room was immediately over the bar, so I had a few sleepless hours until closing time. Over dinner – soup, casseroled chops, roast mutton and the inevitable pumpkin, apple-pie and cream – Dick decided he'd take me on up to Rocky in the car and go on to the coast from there for a short holiday. This suited me, of course, so at dawn we crept out of the hotel and set off along the gravel road, stopping for steak and eggs, chips, cabbage and salad, plus a pint of cold milk at the goldmining town of Mt Morgan. It made a good breakfast, believe it or not, though I must admit boiled cabbage doesn't really go with salad. The lettuce wilted visibly.

We crossed over the Tropic of Capricorn just outside Rockhampton, which is a big town with mango trees and bougainvillea and a tropical smell. I thought it exceedingly hot, though the local people said it was cool. Dick saw me onto the train that left at 5.25, and went on his way.

Wendouree, Alpha

27 April

Well, here we are then, safe and sound but very tired. The westbound train got in to Alpha at 4 a.m. I didn't know if anyone would be there to meet me at that ungodly hour, as nothing had been confirmed, but Mervyn Carruthers was there with an automatic Mercedes. He's a nice person, if a little melancholy; he'd only been married for two years when his young wife was killed in a car accident while driving south to collect furniture to bring here. Their 11-month-old son Drew was on the back seat of the Mercedes when his mother rolled the car off the dirt road. Merv, following along in the truck, was first on the scene; baby Drew was only bruised, but his wife was dead. What a shocking experience for the poor man.

Drew is staying in Sydney with various aunts, as Merv doesn't intend to bring him up here until he sees how I get on. His old mother, Mrs Carruthers, has chronic arthritis and is confined to a wheelchair that she motors about the house: a modern, prefabricated building, completely gauzed in with flyproof netting and designed to catch every breeze. The station is just within the tropics and the country smells quite different from that I've just come from, but it needs rain badly; feed is getting short.

It doesn't look as though there'll be much riding here. The horses all seem to be descended from brumbies and look pretty rough. Merv tells me I can do the milking for a start, so it will be interesting to see how I get on after my abortive efforts at Potter's Flat.

4 May, approx.

This letter will be a hurried scrawl between musters – a jillaroo's work is never done! I have just returned from bringing in 500 head of assorted shorthorns and Herefords with the two roughriders who are working here for a few days: Don Hafemeister and Doug Spann. They're quite gorgeous and I've fallen for them both! They tour around following the rodeo circuit, working at anything available in between each show. Last year Doug was champion bulldogger of New Zealand, and I believe Don is Australian saddle-bronc champ – or it might be bareback, I'm not sure.

Anyway, I rode a little horse called Blue, who is full of life and a good, fast walker. The dust behind 500 bovines on dry ground is indescribable; on many occasions I couldn't see beyond Blue's ears and could hardly breathe. Doug rode his strawberry roan pony called Twinkle, using a bitless bosal on her and a western saddle with wooden stirrups. She really is a pet, following her master about like a dog and travelling to rodeos in the back of his truck in a specially made stall behind the cab.

Yesterday Merv took Don and me in the big Ford truck to the next-door station, Hobartville, to collect some horses to use for a few weeks. They were duly loaded and had a rough trip back, jolting over tree roots and rocks, and rearing and crashing about in the back. They were allowed to rest for a spell in the yards. They all bucked and rooted; in fact one skewbald mare was so bad that she was named Christine Keeler on the spot.

On the second day of my new job I spent from 8 a.m. to 4 p.m. in the saddle: 'rawhide' wasn't in it! We had lunch at a dam while the cattle rested, and the boys gallantly allowed me to lean against the only tree in the vicinity, seeing I was new and female to boot. My sandwiches were warm and soggy and covered in gritty dust, but they went down very well nevertheless. Don and Doug are the best of mates, laughing and joking all the time as if life's just there to be enjoyed.

On the way home Blue nearly did a vertical take-off before trying to bolt when a large goanna rustled up a tree beside us. It gave me a fright too.

The only indoor work I have done so far is to sweep around, make the beds, help prepare meals and wash up. When Drew comes up from Sydney, however, things will probably change, so I must make the most of any riding work meanwhile.

Yesterday Merv made me drive the Landrover over a rock-strewn creek bed with vertical cliffs and large boulders in 4-wheel drive, saying that if I didn't stall her I could pass my test as far as he was concerned. I did everything else wrongly, but somehow reached the other side without stalling, so I'm now an official Landrover driver.

It has become freezing cold at nights now. At six in the morning, just before sunrise, the temperature drops to around 48°F, which may sound like a Scottish spring day to you but to me is perishing. With the temperature at around 70°, the daylight hours are beautiful, though, sunny and clear.

A flock of emerald-green wild budgerigars flew down to the water troughs yesterday. It's lovely to see them flying freely instead of moping alone in small cages, as most people back home know them. I have also seen lots of grey and pink galahs, and a tiny, pillar-box red-breasted wren with a black head that looked like some plastic decoration from Woolworths.

Monday was a State holiday, and although the boys offered to work, Merv gave them the day off. I was free for a couple of hours in the morning, so spent a most enjoyable time by the big shed with a tin of beef dripping and kerosene, greasing saddles, while the radio from Doug's truck blared forth and little Ned Kelly (their adopted Kelpie pup) savaged my boots. Don played his guitar for a while and generally fooled around. He is a natural clown.

Merv took me out to look for the milkers in the evening in the Landrover. I drove back slowly behind the cows and their calves, and then Merv hopped out to pen up the calves and told me to park the 'rover in the shed next to where Don and Doug sleep on old iron beds. There was a worried chorus of 'Whoa!', 'Stop!', 'Everybody out!' as the two lads leapt off their beds, where they had been reading Donald Duck comics, and vociferously helped me to park the vehicle.

I seem to have caused some jealousy around here, quite unwittingly. When Merv asked me to help with the muster the first morning, Coral, the manager's wife, insisted on coming out the next day. Her husband, Toby, doesn't like her to go out riding too much, though, so there was an 'atmosphere'. I hope it's all straightened out now. After all, riding is part of my job and one of the main reasons I came out to Australia in the first place.

10 May

This week has gone, and so have the two rodeo riders, which is perhaps just as well as I was beginning to fall for Doug rather heavily. He's like a character straight out of a cowboy film, complete with bow legs, and looks awkward when he's off his horse. The two of them left on Friday night for the local rodeo.

On Sunday, the day of Alpha rodeo, Toby and Coral gave me a lift into town to see the show, which was very exciting, especially when Don and Doug were competing. Doug in fact won the steer riding, and came second in the bulldogging, and Don was third on the bareback buckjumpers. Bulldogging, called steer-wrestling in America, is when a rider tries to throw a galloping steer by leaping off a horse, which runs alongside, and rendering the young bullock immobile by tying a front to a back leg with a short length of rope. All this is to be done in the shortest time possible: the record is somewhere around three seconds. Doug's little pony Twinkle is lovely to watch in the ring. After Doug has baled out onto the steer she gallops on under her own steam, then trots back to her master when he whistles. I had a pleasant chat with the boys after the show, and Don has promised to send me the addresses of some of the big stations he worked on up north that employ jillaroos.

There's been a lot of ill-feeling about Doug and Don coming to work here. Toby and his offsider, John, wanted local people to sign on for the muster, so of course found fault with everything the boys did, criticising their way of riding and their American saddles and gear, and hinting that they were no good at anything anyway except riding bucking bulls. I think it's most unfair. To me they were splendid, doing more work with much greater interest in and sympathy for the stock than all the others put together, and really cheering up the place with their high spirits. Or am I biased?

17 May

I have had quite a pleasant week – no riding, but a lot of driving, which is good and necessary practice. One whole day was spent doing three weeks' laundry in the washing machine, so the old lighting plant had to work overtime for most of the day. Of course a frog had decided to roost in the overflow tube and became entangled and mashed up in the white sheets. Not a good start to the day.

The next day I grabbed the chance of helping in the yards, where 200 cows and calves were being drafted. I was given the job of keeping the crush full of calves, as Merv said I don't shout and curse at the youngsters like the men do, thus confusing them. You try yelling in that dust; I could hardly utter, let alone shout.

It's my job to bring in the milkers every other night – still a favourite chore, as the bush really comes to life at that time of day. Masses of birds come to drink at the creek, and also the horses and groups of kangaroos. This evening Mary's little calf played up disgracefully. He ran straight into the pen and out through a small gap in the rails on the other side, so round I went only to find him standing beside a *huge* Hereford bull, looking smugly at me from his safe vantage point. I had to go back and fetch his mother before he decided to comply and go into the calf-pen, and *then* he found a further hole to squeeze through and trotted happily into the horse-yard. By the time I'd blocked up the holes and finally put the three calves to bed it was dark.

Friday was spent washing and polishing the windows inside and out – thirty-three panes, each 3 by 2 feet – which had been put in a year or so ago and had never had the putty marks removed. I saw nothing but windows that night in bed, but both Merv and his mum were so grateful that it was well worth the effort. Merv is going to put up some curtains next week. He hasn't bothered with things like that since his wife died, but he seems at last less depressed, thank goodness. We went for a run in the 'rover this afternoon to see a Mr Fingers, who is droving 500 cows and calves to the railway station at Alpha. As Wendouree is on the public stock route, Merv went along to see that everything was in order, and that the cattle weren't camped on his property and eating too much of his precious grass. The drover has bought over 2000 head so far, at not less than $80 each, and this is just re-stocking his losses after the last drought.

I had an unusual day on Sunday. As it was more or less a day off, Merv suggested we spend the morning exploring his western boundary, which he has never seen before, returning in time for lunch. Off we went in the Landrover – after mending a fence broken in three places by galloping brumbies – along a faint track that soon petered out into wild, virgin bush. The going became extremely rough: nothing but washouts and sheer cliffs and rocks, no grass at all, just rocky ridges with groups of thin trees close together and thick scrub about three feet high. The water in the radiator boiled twice, and we had to stop to let it cool down. Then at midday we had a puncture, by which time we were thoroughly lost. We had travelled in 4-wheel drive most of the time and were never out of second gear, backing and turning and getting boxed in. There was no sign of animal life out there, presumably because of a total lack of water. In fact all we saw in the whole day was one kangaroo and two feral cats. Merv mended the puncture and off we went once more. We only had two apples between us, thinking we'd be back for lunch, but luckily the waterbag was full, even though the 'rover drank most of that.

It was 4.30 when at last we limped home to a very worried Mrs Carruthers. According to the speedometer, we'd done 44 miles in all, and most of that was backwards!

There's a funny old man here just now who traps dingoes for a living. He has a 1926 Chevrolet, the first car ever to go on the Alpha to Jericho mail run, complete with wooden spoked wheels and the original green paint. His name is Paddy – John, who was working here last week, is his son – and he really is a filthy, smelly individual. He throws the dead dingoes on top of his swag and tuckerbox in the back of the car, which is full of strychnine, maggots and decomposing bits of dog. Not my idea of a pleasant occupation – but then, it takes all sorts.

I saw some brumbies today for the first time, two mares and a stallion. Merv shot the big black horse with his .303, but left the others, as a neighbouring grazier has his eye on one of them as the mother of a good foal, if she is served by a suitable stallion. The two mares wouldn't leave their dead leader, though they were very afraid of us and the vehicle. I was terribly upset by the whole thing, and didn't speak to Merv all the way home. However, he managed – just about – to convince me that brumbies are a pest and have to be killed. He reckons there are over 400 wild horses roaming in this district, breaking fences, muddying waterholes and eating twice as much grass as a bullock, so sadly they have to be shot, or captured in droves for dogmeat. What a waste of horseflesh, when you think of the number of people who long to own a horse! Of course a lot of them are inbred and misshapen, with ewe-necks and huge heads, but a good number are fine-looking animals. They don't take kindly to captivity, though, so they are not worth breaking in. A few amenable foals are produced, but not many.

I haven't been paid yet, but as there is nowhere to spend money and nothing to spend it on, there is little point in having any. Actually Merv has told me to ask him if I need some. Otherwise he'll pay me by cheque some day. Sixty-four dollars per month is not too bad to my way of thinking, especially when the work is so enjoyable.

24 May

Great drama here yesterday. Old Paddy the dogger had a stroke before breakfast and fell into his campfire. His son John strolled over, rolling a cigarette, to tell Merv, who promptly decided to take the old man into Jericho by car, with me as passenger to open the ten gates between here and town.

Paddy was sitting up in the back seat talking fairly reasonably when he suddenly had another fit. He groaned horribly and jerked about like a cow with staggers, his eyes rolling back until only the whites were visible. His face was an awful yellow colour and a trickle of blood ran down his chin, which alarmed us very much. We discovered later that he had only bitten his tongue. Anyway, Merv put his foot down and the Merc sped along the rough bush track at about eighty miles an hour, bouncing over the endless potholes and washaways. Meanwhile Paddy was trying to reach for the door-handle to bale out, and I had to control him as best I could from the front seat. Then he grabbed my hand and tried to haul me over, saying 'Come here, girlie!' over and over again. All windows were shut except one, and the heater was full on in an attempt to keep the wretched chap warm, but he'd wet himself and the seat during the fit, so we were nearly passing out with heat and fumes. Merv braked at a roadside homestead, ran up to the front door and asked the woman there to ring through to the bush nurse in Jericho to meet us on the road. It was then the covers fell away from Paddy's body and I saw to my horror that all the skin had been burned off one of his legs.

We raced on, stopping briefly to open the gates, which I left open to save time, taking a chance of some stock straying through. The sight of the stout bush nurse standing waiting for us by her Landrover came as a vast relief to both of us. She congratulated Merv on his speed, took a brief look at our itinerant dogger, injected him with morphine and wrapped him carefully in a blanket. After wasting half an hour in Jericho while all Paddy's relations gathered round the car and discussed at length what to do with him, we drove on to the westbound road towards Barcaldine, from where an ambulance had set off to collect him. Merv flashed his lights at the approaching vehicle some ten miles out of town, the two men in the ambulance transferred Paddy onto a stretcher and loaded him, turned the machine around in a cloud of dust and returned to Barcaldine and a good hospital. Merv and I rewarded ourselves with an ice-cream, then I drove slowly home to let Merv relax after his mercy dash. He shut the gates that were still open; fortunately none of the stock had discovered them.

Good rain fell last Wednesday; actually only 30 points fell at the house, but the storm went right through the centre of the property and filled the waterholes, setting some of the creeks running for a short while. I went down in the Landrover to Doolan's Waterhole and found lots of small dead fish floating on the surface and others gasping for air. The water was full of dust with the sudden downpour and thus short on oxygen. It drizzled all night – a lovely sound, the first rain since January. As there is no garden here, not even a fence round the house, I have started digging over a plot and planting geraniums, lettuce, parsley and tomatoes – but I foresee difficulties in preventing the seedlings being eaten by the milkers and horses.

It's really wintry in the mornings now. Everybody had a long lie-in until 7.30 on Sunday, milking the cows after breakfast for a change, and my breath came 'in white pants', as you so rightly say at home, for there was almost a frost. The household water is heated by the sun shining on a collection of copper pipes on the roof, so of course is beautifully hot on sunny days but cold on cloudy days and at night. Merv intends to connect the system to the generator so that we can have some hot water in the mornings.

One of the horses has got out and into the next-door property. Guess which one! Christine Keeler, of course – the most useless creature on four legs. Merv thinks he'll leave her there and she can join up with a herd of brumbies. It's about all she's good for.

3 June

I have a little 'poddy' (orphan) calf in my charge just now, called Bill. He has become very tame, with me at any rate, and has learned to drink his Denkavit after a lot of sneezing and blowing and spilt milk. He now takes two gallons a day, plus a little lucerne, hay and water and is much more alert and healthy than he was when we first brought him in. His mum must have died out in the bush somewhere.

Sunday was mostly a day off. I spent some time making a birdbath out of a sheet of tin; Merv soldered the corners into place for me. Then I put a layer of clean sand in the bottom and a dead branch sticking up out of the water for the birds to land on. It was set on top of a 44-gallon drum under a small tree. No birds came near it that day, but next morning a whole flock of soldier birds – they look like grey minas – landed right in the bath, knocking the branch out altogether, splashing and yelling like school kids on holiday. Since then only the resident willie wagtail and lousy jacks have come in the evenings.

I have also planted fifty-six sweet peas and ninety-six ordinary peas in newly dug ground, and made a new bed for sunflowers and stocks round the tank stand, while some melon and pumpkin pips await attention. Everything will come up at once, no doubt, and knowing me, it's all bound to get out of control.

Merv took his mum, his cousin Doris, who is staying for a bit, and me for a run in the Landrover to the far western paddock, Greentrees, where we saw a dozen brumbies, including a foal, three kangaroos and lots of cattle. As there wasn't room for four in the front of the 'rover, I volunteered to stand in the open back, leaning on the cab all the way and ducking branches every so often. It was a lovely run, and we all enjoyed it, but Mrs Carruthers has suffered for it, and has been quite ill for the past few days. Today she's a bit better, and Merv talks of taking her back to his sister in Sydney soon. If he does he'll have to employ another woman to keep me company and stop the gossip before it starts. He's also thinking of getting a really good horse for me to ride and look after, so that when I leave he'll have a well-behaved animal for any of his young relations to come and ride.

Oh yes, the climax of the week: your daughter was very nearly demolished by an enormous bullock in the yards on Tuesday. There were three cows and two bullocks in one yard, and I was to draft off the cows into another yard. I was standing waiting at the appropriate gate while Toby moved all five around to face me, when suddenly this great hairy beast lowered his huge head and swung his evil horns straight at me. I only had time to say 'Hoy!' somewhat feebly and leap for the rails, and then the bullock proceeded to knock hell out of the gate. Merv said later there were about three inches between his horns and my backside. The same fellow must have been pretty stirred up all day, because when they were trucked out to their paddock he came flying down the ramp, knocking his companion aside, and galloped away at full speed. What you in England might call 'an ill-bred bovine'; in this part of the world he's known as 'a toey old scrubber'.

10 June

The day before yesterday Merv and Toby killed two sheep bought from a neighbouring station, so now we are having mutton three times a day: chops and eggs for breakfast, cold loin or flap for lunch with salad and a hot roast at night. No doubt we'll start to look a bit sheepish soon. After skinning them they brought the carcases into the kitchen and chopped them up on the white melamine table – what a bloody mess!

It started to rain that night and continued throughout the next day, so everyone is delighted; this is the third lot of rain since my arrival, which makes Merv think I'm a rainmaker in disguise. Alice Springs in the Red Centre had *fog* for the first time in its history, and there are repeated flood warnings for coastal districts.

I am getting the urge to go riding again, but Blue is impossible to catch outside the yard, and it would mean getting all the horses in just to catch him for the occasional ride. Merv has asked the station agent in Alpha to let him know if they find a quiet horse anywhere for sale, but this is highly unlikely, as horses are worth $22 each for dogmeat, and no one can spare the time to gentle them. Lead me to them; I'd certainly try.

18 June

Really, mother, it isn't my house, so I can't very well grow red peppers and sweet corn and chillies all over the place. Anyway, the folks out here don't seem to go in for that kind of oriental herbage. Curry powder comes out of a tin here.

Yesterday we started a new section of fencing out in Greentrees Paddock, which meant driving out there in the Landrover. A big bulldozer was already parked way out in the thick bush miles from anywhere, but her battery was flat so I had to tow her in 4-wheel drive while Merv steered to get her going. While he used the 'dozer to push trees and scrub over to clear a swathe for the new fence, I walked ahead pulling all the rusty wire from the original fence out of the matted grass and burnt posts, rolling it into some sort of order, then walked back to drive the Landrover along the cleared ground to where Merv was working. At one point I had to find a detour up a thickly timbered ridge with a rocky outcrop on top, where the 'rover seemed to sit up and beg, for all she was in low ratio and 4-wheel drive. I must have covered quite a few miles on foot, for I was dead beat last night.

It was Mrs Carruthers's seventy-fifth birthday on Wednesday, and as she has been wanting a chook house for some time we decided to build her one. Merv dug the post-holes, I scooped out the loose dirt, then we went out in the 'rover with the chainsaw and cut posts out of gidyea, which is extremely hard wood. It's hard work, and difficult to find enough straight bits, but we got all the posts into position and firmed in, and are waiting for the wire netting to come with the mail, together with the point-of-lay pullets themselves. Now there are just the nesting boxes to build, and some sort of shelter for the wet weather.

We fenced in the rest of the garden the other day, so that the plants are safe from Peggy, June and Mary, plus calves; also from

Toby's dogs, Bob, Damper, Mandy and Doughboy. Toby and Coral have been away for the Rockhampton Carnival this week, so his father, the manager of next-door Hobartville, has had the dogs. A few days ago the mailman ran over Bob's hindquarters, breaking a thigh and leg. They are thinking of putting him down, but it's taken an awfully long time for anybody to do anything about it. Bob has survived being run over three times before, so presumably he's expected to look after himself again.

Talking of broken bones, Dick Perrett wrote the last mail day and says Mrs Golden has broken her arm. I don't know how she'll manage to feed the chooks, start up the lighting plant, cook and do the hundred and one other things she does daily, but I expect she'll find a way, for she's of true pioneer stock.

25 June

Last Sunday afternoon we went round the water points to check that all was well with the mills, bores and troughs, and just on sunset saw a big dingo coming down to drink. Merv shot him from about two hundred yards away, looking straight into the sun. He was a lovely young dog, and it seemed an awful shame, but they really are a menace.

Toby came fencing with us on Tuesday. He drove the bulldozer while Merv and I walked ahead picking up the old wire and posts. We took a packed lunch, and by the end of the day had cleared about one and a quarter miles. We did the same on Wednesday, so the nine miles of new fence-line is getting on slowly but surely. Merv intends hiring contract fencers to dig the post-holes, cut the posts, drill them and string the three strands of wire along, as it's too much of a job without the right tools. The fence is being built around a few hundred acres of land, where a poisonous bush, 'heartleaf', is growing, to cut the area off from the rest of Greentrees Paddock. Greentrees contains good lush feed on which Merv hopes to fatten his bullocks later. Heartleaf is deadly after rain, for fresh green shoots appear that cattle love to browse on – and of course they're dead a few hours later.

Merv and I had quite an eventful trip into Clermont, a town about one hundred miles away, during which we were rammed by a kangaroo and had a blowout. The poor old Merc's cream door is badly dented, though the 'roo was apparently unhurt and bounded away shaking his head a little. They're funny animals in that they head straight for a car. Either they don't see it coming or they're stone deaf. We did some shopping in the prosperous little town, had lunch in a small cafe and wended our way homewards towards evening.

A young honeymoon couple are staying here for the weekend, and yesterday we took them for a run round the property, stopping for lunch under a blackbutt tree by Greentrees Dam: tinned salmon sandwiches, apples and billy tea. The couple were thrilled to see first nine brumbies running, then a mob of six that came cantering right up to the 'rover before veering off. They were led by a massive red-roan stallion with a black, tangled mane and tail, at which I pointed my Brownie Box camera. We were jolting along at the time, however, so I don't suppose it will be much of a picture.

2 July

The honeymooners left early on Monday morning, and after they had gone Merv sent Toby out alone to clear some more fence-line while he and I tidied out the big shed. We put all the hay and bags of feed at one end, tying it off from the milkers and horses, cleared out another section to park the cattle float in out of the sun, and burned the huge heap of accumulated rubbish: tyres, old tea-chests, moth-eaten saddles and endless bits and pieces. He lit the heap while there was no wind, but when the bonfire was burning away merrily a stiff breeze sprang up, sweeping the flames into the surrounding grass. Merv rushed around with a rake beating at the burning grass while I tore off to fetch wet sacks. The fire could easily have got out of hand, as the grass is yellow and bone dry, but all was well.

On Tuesday we all went out fencing – and finished it. Jubilation! The tractor was then turned around to start clearing a further width to form the basis of a good all-weather track alongside the new fence. A small herd of brumbies came galloping past, bucking and squealing and looking really beautiful. There must be masses of them out there; we've never seen the same lot twice.

On Wednesday Merv took me into Alpha. There was a strong wind blowing and the little town was a real dust bowl, with sand, leaves and litter blowing across the railway line and along the main street. There was nobody about, although the shops were open; it looked like a ghost town of the Old West.

Yesterday was spent in the garden digging holes for six

grapevines, six citrus trees, six rosebushes and two passionfruit vines – all of which are due out from Rockhampton in the mail today. The beans are rampant, the carrots, beetroot and peas look green and healthy, and the sunflowers seem to shoot up inches in the night.

It was announced over the radio that Rodney Perrett has been balloted a cattle station in central Queensland somewhere, so I must find out more about it. I'll bet he's thrilled. Not many people own their own stations; most of them are leased by the Government for thirty years provided one pays a minimum sum on improvements like fencing, water points, new yards, etc. At Rodney's age it won't matter if he is in debt for ten years or so, because by the time he is 30 he should be pretty wealthy. It also means he won't have to go to Vietnam, thank the Lord. There were 143 entrants for the block; Rodney was just lucky to be the first name out of the hat.

Poor old June, one of the milkers, is sick. She has rheumatism in her offside hind leg and looks thoroughly hunched up and miserable. Old Peggy with the curly horns has been knocking her around when I feed them at night, so I keep June's hay separate now. Actually it's high time her big fat calf was weaned.

Friday night – 8 July?

You should have received a parcel containing a book called *Brigalow* and my latest films by now. I loved the author's style and can't think of a better book to give you a good idea of mustering and stockwork, even though it is set in the pre-war years. Things haven't changed much in the cattle country.

Well, we haven't been away from the house much this week – only once, in fact, when Coral and I went to Surbiton station to get the two killing sheep. We set off in the 'rover, me driving, but the thing started making terrifying noises a couple of miles out, so I drove back in fear and trepidation. Merv said it sounded as if there was a tooth off one of the gears, so we set off again in Toby's Holden, with Coral driving this time. The sheep were put in the boot with three legs tied together, and Coral opened the lid at every gate on the way back to let them breathe.

Merv and Toby have been spending the week erecting the stands at the back of the house for the rainwater tanks, doing a pretty good job with the new welder Merv bought the other day. Consequently I have been doing lots of gardening, though today, after giving the house a quick once-over, I managed to grease four bridles and one saddle. Merv's uncle has put a deposit on a grey gelding, and is going to get two or three other horses for us – whoopee! However, before they arrive we'll have to divide the milkers' paddock into two to keep the horses separate from Toby's, and that means running water in from the house, buying a new gate, etc., all of which involves more expense. Also Merv has decided to go down to Sydney on Tuesday to collect Drew, and maybe a puppy to keep him company, and new springs for the Mercedes . . . I can't see myself having much time to look after a child on top of everything else, but no doubt I shall have to manage. Most days I have an hour or so to myself after lunch, and today I took my camera and James Bond down to the creek, sat under a tree and read. Lovely, except for the flies; the horses and hordes of little birds came to investigate.

On Thursday eight 'strangers' (neighbours' cattle) got through the fence into the milkers' paddock, so I walked right along beside the fence with a pair of pliers (talk about boundary walking!) and patched up the loose wires. Two bulls were standing on the other side; it must have been they who pushed the wires apart with their big shoulders. Toby will have to get the strangers out, but how he's going to catch his evasive horses with the Landrover out of action, heaven knows. Still, that's his problem. I could have done with a racehorse yesterday while bringing the milkers up. Two of the aforesaid strangers came along. Mary's little calf shot off to join them, his mother stopped and bellowed a reprimand, Peggy chased another stranger in a different direction, so there were cows everywhere.

The temperature was down to 32°F this morning before dawn – the third frost of the winter, and by golly it was cold. Luckily I covered the sunflowers, tomatoes and beans with old saddle blankets before dark last night.

There's such a lot to be done this weekend: finish the chook run, find nesting boxes, plant the citrus trees, lay the tiles in the kitchen, do an enormous dhobi, put up a bed in Merv's room for Drew, see a man about putting a telephone line in . . . Poor Merv reckons he'll have to sell at least eighty head of bullocks at $100 each to make a bit of money, and he'll be lucky if he can assemble that number without including some of the nice young steers.

I scalded my arm on the kettle the other day, but it's okay now. That reminds me, the Barcaldine doctor said Paddy will be in hospital for at least another two months. They were going to amputate his leg, as it was burnt to the bone, but they haven't done so yet. It seems he's had another fit too.

15 July

Merv flew off to Sydney on Tuesday to collect Drew, leaving me with Mrs Carruthers – and of course the horses arrived on the goods train yesterday. The wire came on Wednesday, so Mrs Carruthers told me to go in with Toby and Coral and keep an eye on things, as she doesn't trust them further than she could fling a herd of elephants. I was dreading the trip into town. I fear the antipathy between Coral and myself is mutual.

Anyway, we set off, the cab of the float reeking of cheap scent and hair lacquer. On arrival in Alpha we found we had a slow puncture, which Toby had to fix while Coral and I sat in the cab in stony silence. Finally she began saying how stupid it was of Merv to bush all Toby's horses: why couldn't they run together with the new ones, etc., etc. I carefully explained that we wanted some quiet horses that weren't descended from brumbies, as neither Merv nor I were particularly good riders; also that the horses would be kicked to bits, being new, and anyway none of Toby's horses were reliable. Coral was most put out at this, and relapsed into offended silence until the tyre was mended.

The horses duly arrived on the train, and after much shunting were unloaded into the receiving yards. Toby collected the waybill and permit, which Coral read as we walked back to the truck. In all innocence I asked if I could have a look at it after her, and she turned on me any yelled: 'Who the bloody hell d'you think you are to come out here and organise everything? Toby's the manager, though you wouldn't think it! Let him have a say for once . . .' blah blah blah. I was horrified, and shouted back at her to stop being so utterly childish. The whole of Alpha must have heard us. Toby kept mum; he obviously didn't dare interfere between two warring females! It was all so unreasonable and infuriating, and from then on we were like a couple of hostile dogs, all stiff legs and raised hackles! Anyway, Toby and Coral deliberately wasted hours drinking in the pub while the horses stood in the truck. We didn't set off for home until five o'clock, and I knew it would be hopeless to unload them in the dark in a strange place, but that precious pair didn't care. After all, they were not *their* horses – neither were they mine, but there's the difference. I read a magazine all the way back and got out to open all the gates. Of course madam with her nylons and hairspray couldn't possibly jump down into the dust to perform such a menial task. With any luck I might have spoilt her 'day in town' for her, which is some consolation. Am I being *too* bitchy?

Anyway, we put the poor wretched animals in the night-paddock in the dark, and I had to find a rug for the grey who, being clipped, and coming from the humid coast to this penetrating cold climate, would suffer at night. Mrs Carruthers said to use the blanket off my bed, so I strapped it on tightly with a surcingle. It must have slipped back, though, because this morning it was torn to shreds and thick with grass seeds.

Today we spent several exasperating hours trying to catch the newcomers. There is a chestnut mare, a black mare, two bay geldings – one a thickset fellow who looks as if he was a stallion until very recently – and the grey gelding. First I tried to get them to come to hand; the lovely chestnut mare came up, grabbed a mouthful of hay and dashed off. Then I attempted to walk them up the fence to the milkers' yard, but they wheeled off and raced to the other side, stopping there to see the effect it had on me. My language was choice, but it didn't help at all, and after one and a half hours I gave up, taking their Riverina stock feed down from the shed and feeding them in the night-paddock. I was hoping to catch them and take them up to the house to show Mrs Carruthers, who of course can't get out of her wheelchair. In the afternoon I tried again, this time managing to catch the chestnut and tie her up in the yard. Then I tried to drive the others up to her, and they were almost in the yard when the mare pulled away from the rails and galloped off with the others in hot pursuit. I was left behind a tree, full of dust and fury. I went home and cut my hair as a result. What makes me like horses?

The chooks laid four eggs on their first day – from fright, I guess, for since then they haven't produced a thing. I caught a lousy jack in there today; he opened his beak and screamed like a hare in a trap. I should have wrung his neck – the wretched birds have already pecked off two sunflowers and some beetroot – but couldn't bring myself to do it. A couple of the prettiest little yellow and black birds with pink and white cheeks and chopped-off tails – like tiny Indian pittas – have dug a tunnel in the side of the hole intended for the mulberry tree, presumably to nest. They are very tame, and don't mind me working near by.

My turn to milk tomorrow. I must remember to take a pair of scissors along, because Mary's udder is very hairy and thus difficult to handle. On that note I shall leave you for this week.

23 July

You ask if I'm unhappy working in such heat, to which the answer is no. I'm not fed up with the sun at all; in fact it isn't hot enough just now. All this week we've had frost, 27° F being the lowest temperature. All the latest beans have been nipped and the mornings are *perishing* at 6.30, when we get up. The sun rises at about 7 a.m. and doesn't warm the air until 9.30 or so. No, I don't want to come home yet. I really love this life, for all it's hard work. In fact, I don't think of it as work, because I enjoy it so much. I can't see myself staying in Australia for ever, though, and will be pleased to come back home at the end of two years.

On Saturday afternoon Toby had a violent quarrel with poor old Mrs Carruthers. (Merv was still in Sydney.) She told him to put all the new horses into the milkers' paddock, which he did, but he kept the grey in and put hobbles on him so that he could be caught easily. Mrs Carruthers sent me down to take them off, so Toby raged at me first – understandably, perhaps – then strode up to the house, called the old lady a bloody interfering old bitch, said he refused to work for two women any more and that he was quitting. Soon they were both shouting, but as Mrs Carruthers courageously held her own I kept out of it; otherwise I might have struck Toby with something heavy.

Anyway, he and Coral are packing up to go, thank the Lord! Poor Mrs Carruthers has been in a state of nerves since the row, and so have I – though I think in my case it's more likely supressed rage. We locked the doors that night for the first time in Wendouree's history, for fear of 'retaliation'.

On Sunday I went out to see the horses, but they fled. On Monday I milked and got twice as much as before – almost two gallons. I tried again to approach the horses, with feed, but no luck.

Next day Merv came back with Drew, a handsome little boy but rather spoilt, not unnaturally. I walked two miles out to the boundary and painted the gate and posts white, and the car came along just as the last drop of paint went on, which was the idea of course, saving me the walk back. Merv was delighted.

Breakfast next day was chaotic. Drew won't eat anything and is so dependent on his dad that he gets hysterical if Merv leaves the room for a second. I played with him that first day, fixing up a saddle on a petrol drum and tying a rope to a tree in front of it so that he could play cowboys and Indians, but his heart wasn't in it, poor little kid.

I caught the black mare that afternoon, my favourite of the new bunch and called Twinkle after Doug's pony, and started to lead her back, but she reared up and pulled away, the strap still round her neck. However, next day I took some bread to the horses, and surprisingly the grey and Twinkle came straight up and ate some, though perhaps they knew they were not going to be caught.

This morning at 2 a.m. I had to wake Merv (being careful not to disturb Drew, who sleeps in his room), and we set off for Alpha to fetch Uncle Archie, who was arriving on the Midlander at 4 a.m. Uncle Archie is the bloke who bought the horses for Merv and is to stay indefinitely: a tough, wiry old fellow who talks about nothing but horses and declares he owns a racing stable, which I doubt.

We got back at 5.30, and were having breakfast when Drew woke up yelling. Then the four of us set out in the 'rover to get a piece from the tractor at Clara's Dam, and to show Uncle the property at the same time. We did a circuit of the place and just inside Doolan's saw some 'roos, which we stupidly chased, coming to a sickening halt with the front wheels facing in different directions: all four axle bolts had broken on a stump. Merv took the wheel to pieces, put the bits in a bucket and gave it to Uncle to carry, I took the waterbag, he carried Drew and we started to walk the six miles home. We began at eleven and arrived home surprisingly fresh at a quarter past one. I'm tired now though, so will go to sleep.

Oh yes, we ran over a dingo on the way to town last night, just inside the boundary.

30 July

Glad to hear you are enjoying *Brigalow*. Really it could be about any of the properties I've been on so far.

This has been an even busier week than usual, what with Drew, the horses, Uncle Archie, the chooks, cattle, the house, etc., but it's all good fun, though Uncle is rather trying at times. He's very deaf, has only one eye, talks nothing but racing and every second word is the Great Australian Adjective. Drew summed him up rather well the other day. We were all seated round the table, Uncle was holding forth and everyone was getting glassy-eyed when Drew remarked, 'Bloody horses!'. Unfortunately the old man didn't hear, though he did wonder why everybody laughed.

Drew has become much more friendly and seems to be getting quite fond of me. Yesterday Merv and Archie went out all day to mend the boundary fence, and we expected hysterics and tears, but he stopped crying before his dad was even out of sight. He let me dress him and ate his meals very well – usually no one but Merv can feed him. He's quite a sweet little boy, but I don't seem to have the endless patience required to look after him. So far I've played dogs, kittens, monkeys and horses, and he very nearly talked me into singing him to sleep the other night.

Yesterday I caught Jim, the grey gelding, saddled him and went all by myself into Doolan's Paddock to bring nine beasts plus two bulls through the home paddock to water, and then out into Horse-shoe Paddock beyond. There was a spot of trouble crossing the creek, as the bulls decided to have a scrap, pawing the dust and bellowing curses at each other. I nipped it in the bud by charging Jim at them, scattering the rest of the beasts for miles around and having to round them up again.

Last Saturday, after the 6-mile trek home at midday, I'd had ten minutes' sleep after writing you when that wretched old man woke me up. He wanted me to help him catch the horses. We caught Jim, Flicka and Twinkle; then Nuggett, the solid bay with a bad temper, decided to stir things up a little. He wandered into the middle of the group, bit Jim and kicked Twinkle, then ran off with his tail jammed down tight like a naughty dog. Jim reared up and galloped off with his bridle on, spraining Uncle's finger. It was a most belligerent old fellow who saddled Twinkle and went after Jim, while I led Flicka, the chestnut, to the night-paddock and shut her in, only to discover that the top entrance was open. Obviously she knew that, because she galloped along the fence and out with a triumphant squeal.

On Monday we went into town to get the bit for the 'rover drilled, so that was a day off for me; then on Tuesday the three of us – Uncle, Merv and I – saddled up and rode out to the Landrover with the part. I rode Billycan, the other bay gelding who is very quiet but shies at anthills and pigeons, and we left Merv at the 'rover while we led Jim back home. I rode Billycan again the next day when Uncle and I got the strangers out of a paddock and drove them 6 miles to Doolan's Waterhole. We had great fun cutting the bull away from Mary, on whom he is rather keen just now.

I received an electric shock from Drew's bed the other day. I stripped back the blanket rapidly and the current must have run up the iron bedstead; it shot up my arm with surprising force. It must be the dry, cold atmosphere here. Unfortunately the recent frost has knocked a lot of condition off the cattle; they don't do well in the cold.

Mrs Perrett wrote from Kabunga West this week, saying Rodney's block is not all that good. It has a lot of poor country in it, and they are all a bit disappointed, but she says he's going on with it. They are all going there, staying in a tent, in the August holidays to settle him in by helping build a shed first. There's nothing like starting from scratch.

I caught five butcherbirds and a crow in the chook house yesterday. Should have killed them all, but of course just put them back out through the wire door. Today the young roosters arrive – 4 a.m. cock-a-doodle-doo chorus from now on, I suppose.

Today we went down to Doolan's again, this time in the repaired Landrover, and found the fence between it and Boylan's Paddock down. Uncle and Merv fixed that while Drew dragged me off to play in the creek and dig holes in the sand. I found a dead galah, so now have lots of lovely pink feathers for my collection.

6 *August*

This will be a rushed letter and therefore somewhat jumbled, but I am snatching moments between fixing the two huge rainwater tanks up onto the stand – at least I'm helping Merv and Archie. It's very cloudy, and of course if it does rain today the tanks won't be up in time. I've been driving the 'rover, pulling the rope on the block, looking backwards all the time. Just as well it's a Big Country with nothing much to run into in the home paddock.

Next Saturday Merv's late wife's mother and her son of 13 are coming up, so I volunteered to move out and sleep in the cottage. This house is all very well, but it's all modern and anti-fly gauzed, whereas the cottage has half-inch cracks between the floorboards. All the birds come down to the tank in the mornings for a quick splash and a drink, and the general atmosphere is much more in keeping with my ideas of bush life. Also one hears every sound in the house: Drew has bad dreams still, poor lad, Merv gets up for the occasional drink of milk to ease his ulcer, his mother wheels herself to the lavatory a couple of times each night and Uncle snores like an old band saw.

On Sunday I was looking forward to the usual lie-in, but Uncle Archie got up at 5.30 to do the milking. He just can't seem to stay in bed, even though he's about 77, but he does do all the milking and odd jobs, which is a big help.

Next day Drew was feverish and difficult to deal with all day while the men were out; then, just as tea was ready and the Landrover drove up, he was sick all over the floor. He had a temperature the following day, but is better now.

When I went to collect the milkers the horses came up behind me and decided to liven things up a bit by chasing the cows. Poor old Peggy, who has never moved at more than a slow amble in her life – except when she feels like charging somebody – was galloping flat out, udders swinging, pursued by Nuggett who was doing his damnedest to nip her bony backside. Then Flicka and Billycan camp-drafted Mary away from the others and they all disappeared, leaving me once more standing in the dust with arms akimbo.

Wednesday we had plain turkey for lunch. It tastes like rabbit and makes lovely jelly. I believe they are protected out here, but this one was inadvertently run over, so was eaten instead of being chucked out. Before lunch I went to collect the two killing sheep from Surbiton, 12 miles away over very bumpy tracks. Archie and Drew came too. We tied the sheeps' legs, heaved them into the back and came home again; then as we untied them in the yard one bounded out and gave little Drew an awful fright. He had nightmares that night.

Yesterday Merv and I set out for Jericho in the truck to collect two Wendouree bullocks and a heifer that had appeared on a property near the town. They had wandered a long way from home, nearly fifty miles in fact, negotiating countless fences on the way. One wonders how they managed to get there.

There was no muffler on the truck, so neither of us could hear ourselves think, and the bumps were terrible. I hit my head on the metal roof at least a dozen times, but Merv was all right with the steering wheel to lean on. We reached the station to find there was no one about. There were two horses in the yards, though, so we waited around for three hours, then returned to the town, where we met the owner. He said, very brusquely I thought, 'Take a saddle and bridle and catch a horse and bring them in yourself; there are no men available'. Back we went to the yards, and Merv decided to drive through the paddock first and locate the cattle, as he doesn't fancy horse-work and wouldn't let me go alone on a strange horse. We drove around until we found them, got out and tried to walk the cattle back to the yards, ending up running and stumbling and getting nowhere fast because there was a very toey blue heifer there who did everything at the double, including jumping two fences.

Eventually a boy on a horse came to our rescue and yarded the cattle. Just as I had put the sliprails up there was a combined yell from the boy and Merv. I looked round to see this same blue beast coming flat out for the gate, and fled just in time; she sailed through the rails, breaking the top one, her head up and tail straight as a pole – the cloven-hoofed showjumping champion of the year. So we loaded our three and headed home at last. We'd set out for Jericho at 7.15 a.m., and arrived back at 7.15 p.m., dead beat, evil-tempered and filthy with dust and sweat.

They've got the tanks up I see, and still it hasn't rained.

12 August – Happy Grouse Shooting!

Further to my last letter, the tanks were fixed just in time. Five minutes later the rain came down in torrents; there were black clouds everywhere and grumbling thunder with pink lightning. I joined in the mad rush to get pipes fixed to the gutters, etc., and also had to find a tin roof to shelter the new roosters, which were beginning to look worried. I was like a drowned rat within seconds, but it was lovely to see all that rain going into the tanks instead of wasting away under the house.

In fact, it rained for two and a half days; three good inches fell, and the tanks collected about 2000 gallons. They have really been christened in style. The waterholes are full, spreading the cattle out away from the one watering point in the breeders' paddock to new feed. The home creek ran for the first time since December, and already new grass is showing. The garden has burst into life. I picked fourteen beans and am thinning the carrots.

While it rained we were all confined to the house, and Merv decided to lay the tiles in the kitchen. This proved a long and tedious job, with Drew getting into everything, running off with nails and tiles and being shouted at by all and sundry. I was glad to get out to fetch the milkers and feed the chooks, despite the sodden ground. George, Mary's bull calf, hates going into his pen now and runs in every direction but the right one until the last possible moment. Today he galloped off and took shelter beside the biggest rangy bullock that came from Jericho last week, looking very smug. I don't trust that scrubber at all. He has horns bigger than a Highland cow's, which he waggles menacingly at everyone, and he travels everywhere at a nervous trot.

Nuggett, the fat bay who kicks, fell over yesterday while trying to kick the others and couldn't get up again as he was jammed against the feedbox. I'm afraid I unfeelingly laughed at his discomfort; it served him right, the narky old beggar. Merv had to remove the box before he righted himself.

Alpha has a 67-year-old doctor as from today. He spent twenty-five years in Ceylon, believe it or not, after studying tropical medicine in London. There must be something far wrong with him to end up out here. Alpha hasn't had a resident doc. for many years now, so it will be interesting to see how long he lasts.

Tonight Merv and I are going into Alpha to collect his mum-in-law and son off the Midlander at 3.50 a.m. – ugh! It's still bitterly cold at that awful hour. I'm just about asleep as I write. Forgive me, it's the first time I've sat down today except for meals. Mrs Carruthers seems to have handed everything over to me. The only drawback about staying at a place for more than three months is that one can't help becoming very much involved with people's problems and worries, and Mrs Carruthers appears to think I'll be here for ever. She says she wouldn't know what to do if I left, and even though

I told her quite firmly that I shall be going in November, she wouldn't believe me. Perhaps it's time I wrote to the addresses Don Hafemeister sent me and started to look for another job.

20 August

I have had a very pleasant week sleeping on the verandah of the old cottage. Jean Swinbourne and Terry the mop-haired 13-year-old are such nice people, full of life and great fun. Terry of course wanted to sleep out too, so we carted his bed down and set it up at the foot of mine – there was just room on the small verandah. There was a frost on the first night, but I was wrapped up in Merv's double eiderdown and the blanket I put on the grey horse (it's still full of spear-grass seeds), and Terry had the canvas horse rug.

On Sunday Terry needed little convincing that he wanted a ride, so I caught Billycan without any trouble, saddled him and drove Flicka and Nugget into the yards. I tried to approach Nugget on foot, but he whipped his rump round towards me all the time, his ears flat back and one evil eye watching my every move. He won that round, because I then went out with a large handful of feed to try and catch Twinkle. She grabbed the hay and ducked out of reach, and I had to return to Billycan and drive Twinkle and Jim into the yards. However, when Terry opened the gate to let us in, Flicka raced out, taking the others with her. Billycan started going backwards and then bucked, and we gave up altogether when we found that Nugget, who was left in the yard, had been picking up the sliprail in his teeth and dropping it until it fell down, leaving him free to bolt out as well. Horses . . .

Monday was more successful. I milked – getting far more than Archie ever does, ho-ho – then Merv and the others went into town, leaving Terry and me to muster the few bullocks and heifers in the home paddock. We walked all the horses quietly into the big cattle yards, drafted off Twinkle and Billycan, bushed the others, saddled up successfully and set off, Terry on the bay and I on Twinkle. No one has ridden her since her first day here, and I expected a few bucks, but she only shied twice. She's a slim but strong mare with a lovely stride and an armchair canter, and she loves her work; as soon as she sees cattle her head goes up and she begins to prance. We got all the cattle, including the milkers, much to Peggy's horror. She stood and bellowed loud protests against the indignity of being mustered with the common herd. One bullock got his hips stuck in a gate when we shut it too quickly, which understandably made him rather touchy, but otherwise the mission was uneventful. Having put them in their respective paddocks we went for a gentle ride, ending up with a race down to the windmill. For the first time in my life I galloped really fast (intentionally), but Twinkle is so biddable and easily controlled that she is a delight to ride.

For the next four days it rained on and off, giving us a further 4 inches. Horseshoe Creek is 'running a banker', roaring along across the road. I'm writing this letter seated on a dead tree by the crossing, which Terry has just waded across; the water is waist high. The mailman won't be out today, but I'll continue with this anyway.

On Tuesday Merv and Archie, Terry and I left Jean in charge of Drew and Mrs Carruthers and went out to Clara's Dam to fix the tractor. It poured with rain all the time, but to our surprise the Landrover didn't get bogged. I managed to make a smoky little fire that just boiled the billy before fizzling out, and we each had a soggy sandwich. After that enlivening lunch we started clearing a new track for the fencers to work along, I driving the 'rover and skidding up to trees in a decidedly drunken fashion on the greasy surface, but coming to no harm, while the tractor nearly became stuck twice.

The next day Merv should have picked up the sheep, but it was far too wet on the black-soil plains near Surbiton to drive a vehicle, so eventually he decided that Terry and I should ride over and cancel the sheep for this time. (There's no phone, remember.) It took us over an hour to catch and saddle Nuggett and Twinkle, and it was two o'clock when we set off, hoping to be back by five. We trotted where possible, actually for about 200 yards, otherwise we just plodded along in the mud and reached the Surbiton homestead at five o'clock, all muddied nearly up to our necks. Poor Twinkle hated it, sinking up to her hocks most of the time. The Surbiton folk were amazed to see us and said they weren't bothering with the sheep anyway because of the wet. They lent us a torch and a couple of woolly jerseys to wear on the way home; we altered the saddles on refolded blankets and turned back. It was dark long before we reached our boundary, and the horses were consequently very jumpy, so we started singing to soothe their nerves. We sang every song we could think of at the tops of our voices, horribly out of tune, I fear, but the horses seemed to like it and plodded on with one ear forward and one back, listening. At the creek we dismounted and led them for a spell. It was now seven o'clock and raining again. The little torch helped us to open the gates, but Twinkle shied at the light, kicking up the water behind me all over my bottom and back, so we rode without it. We were almost home when we saw the lights of the Landrover coming to look for us. Jean and Merv had been most anxious because we were so late, but were glad to find us in good spirits and none the worse for our abortive trip. We got home at nine and tended to the horses first, giving them good big feeds and a bit of a clean-up. Poor Twinkle was almost dropping; in fact her legs were trembling. Twenty-four miles of soft mud is more than enough for any horse.

Merv went out next morning and shot a mickey, which we butchered next morning, so now we have rump steak for breakfast, stew or curry for lunch and a roast at night. Suits me; it's relief to get off mutton.

Jean Swinbourne is a real livewire, a very kind and cheerful person. She has pressed me to spend Christmas with them at Maryborough. They have a boat, and the scenery is wonderful, apparently, so I might well accept the invitation. Terry has three older brothers too.

Heaven knows how I'm going to get the cows in tonight with the creek running 4 feet deep and 2 feet of mud below. There's no way across it unless I do a Tarzan act and swing over on a branch! The rain has fallen all over the drought-stricken part of Queensland. It couldn't have come at a better time; now there'll be good feed and water until the December rains.

I'm being attacked by a vast bulldog ant, so must make a rapid move.

27 August

We went out to have a picnic last week and got well and truly bogged. It took three attempts to get the 'rover out again, with everyone collecting branches and anthills to shove under the wheels. The picnic at Rocky Crossing was somewhat delayed, but it was worth waiting for; we sat by a little waterfall, boiled the billy and ate damper and tomatoes and everybody enjoyed it hugely, especially Drew.

One afternoon Terry and I decided to go swimming. We had changed and started walking to the creek in Doolan's Paddock when Merv came tearing along in the 'rover with Jean and picked us up. We swam, while they lay on the bank and watched. It was a nice swim, although the water was rather muddy, and I cut my leg on a submerged branch, inflicting an eight-inch wound, which fortunately isn't deep. It was lateish in the evening when we drove home, so I went out for the milkers still wearing my swimsuit. This apparently disturbed old Uncle, for he told Merv that if he'd still been able to run he'd have been heading in the direction of 'that bloody jillaroo'! And he's only got one eye!

On Tuesday Terry, Uncle and I mustered Boylan's Paddock, where the poisonous bush grows and where there shouldn't be any cattle. Surprisingly we mustered eighty head before lunch, which we had by the bore. I tied Twinkle to a tree and sat on a log near her. Suddenly her head came over my shoulder and she gently but firmly grabbed half my sandwich. I ate only the meat, then, while she had all the bread and butter and pepper. She sneezed once, but seemed to enjoy the flavour.

Merv has been clearing the scrub from around the house with his tractor, preparing to make a paddock where he can keep some killing sheep instead of trailing over every fortnight to collect them from Surbiton. We are still living on the mickey. It's the best meat I've ever had, but unfortunately the small fridge defrosted yesterday and some parts have gone a bit green. However, they will still make good curry.

The cold weather has returned after the unseasonal muggy rain. Once again the nights are clear and frosty and the days fairly chilly unless one is out of the wind. The grass has shot up, Mary is giving twice as much milk as a result, and we had our first vegetables from the garden last night; beans, and a dish of young carrots in white sauce. The peas are swelling nicely. Jim the grey got in during the day we were out, ate the tops off all the carrots and lettuce and trod on the new beans. Why can't people learn to shut gates?

6 September

I've been mustering, so must apologise for not writing on Saturday as usual. You'll understand why when I tell you I've been up at 5.45 every day for over a week, out all day until dusk, then had to bring in the milkers, get the tea and clear up, and therefore have been dying in bed at 8.30. Actually, after the first two days it wasn't too bad.

Everyone except me went into town on Thursday to inspect the new Ceylonese doctor, who speaks indifferent English and apparently has only one drug in the hospital, namely penicillin. Mrs Carruthers has to go in this week for observation, as she has been pretty ill and in a lot of pain lately. I don't envy her.

On Friday the three Wendouree stockmen mustered again, taking the bread and meat out to the fencers first. They are starting the 9 miles of fence we cleared, and have already cut all the posts and laid them out. The following day we saddled up, but Jim was very lame in two feet and could hardly walk, so after stumbling for a couple of miles Merv decided to go back. Not long after this, Twinkle got the scours and weakened rapidly, not even summoning the energy to trot. When we stopped she was panting like a grampus, and Uncle told me to take her back too. I led her the two and a half miles to the dam, stopping every few yards to give her a rest. Her head was propped on my shoulder most of the way, poor old duck. At Clara's I found Jim lying down, so unsaddled Twinkle and left her with him, feeling most inadequate – but what else could one do for them? Merv and I then went on and started to grade the track along the fence, for want of something better to do. We saw two black cockatoos fly over: enormous, rather sinister-looking birds with a vivid red band across their tails that shows clearly when they are flying.

Coming home that night Merv shot two brumbies, but they took so long to drop that I got really upset and couldn't speak for a while. One was a young black stallion and the other a funny-looking, in-bred sort of creature.

On Sunday Nuggett was loaded into the truck, alone in all his glory, and driven out by Merv, while I drove the 'rover to the dam. Jim was still very stiff once he was saddled, and Twinkle was hardly able to move. All four legs were swollen, but Uncle said it wouldn't hurt to ride her gently, so we all mounted and started to cross the creek. Billycan decided he didn't want to, Merv dug his spurs in and the bay reared up and spun round, depositing his rider bottom first into the water – wasn't he mad! I couldn't help laughing at the indignant expression on his face, and even old Uncle grinned faintly – the first time I've ever seen him smile. Poor Merv dried himself off as best he could, and we went on and got the cattle out of the holding paddock – about eighty head. We started to drove them home, but dear Twinkle nearly fell over twice and just couldn't keep up, try as she might, so I went back and left her once more at the dam, got into the 'rover and caught up with the mob to hold the tail position quite successfully. But I'd much rather go mustering on a horse.

Monday was spent in the yards drafting, branding, castrating, weaning and earmarking, the latter being my job. It was very unpleasant, as some of the older mickeys have such tough ears to cut through. Merv did the castrating for the first time, and although he made them bleed to begin with, soon became quite adept at the operation. Later on that evening we tore out to Clara's Dam in the 'rover, loaded Twinkle and Jim into the truck and brought them home. They are still very lame and weak. They must have eaten some poisonous weed in the holding paddock, or perhaps it was a virus infection.

Yesterday Uncle and I took the cows and young calves back to Charlemont Paddock. I was on Billycan, who decided he wanted a drink and walked right into the deepest hole he could find, filling my boots with water. Still, at least he didn't give me the same treatment as Merv got last time out!

Today I am at home and must clean the whole house, which is naturally pretty grubby after more than a week's neglect. The garden has suffered too, but there are masses of lovely beans and four small tomatoes, and one sunflower is about to bloom.

12 September

Just a short letter this week, as there's nothing startling to relate, except that Merv's mum is now in Alpha hospital, poor soul. On Thursday we left for town to deposit her in the hospital and to see Uncle Archie onto the plane. The airstrip is tiny, with a shed, a pile of dingo bait boxes and a lot of nothing else. We waved goodbye to the old man as he boarded a Fokker Friendship, then took his sister round to her bed in the tiny cottage hospital, where we had to wait an hour before the doctor deigned to come out and see her. He is apparently unpopular; already the matron has quit and the nurses are rumbling with discontent. There was a pleasant sort of pi-dog wandering in and out of the small wards wearing a charming grin on his lopsided face. He seemed to belong to the place, for he flopped down in the sunshine on the verandah and nobody booted him off.

On Saturday we went in to see Mrs Carruthers. She's not at all happy; in fact she hates being there, but has to stay until her tests are completed. What sort of tests we never found out but can only hope that 'they' know best.

I saw Twinkle yesterday. She's still very lame, as if she's worn down all four hoofs to the quick, which she hasn't. It took her ages to hobble over for a biscuit. Merv has put back the weaning muster until October when Uncle is coming back to give us a hand, so Doug and Don won't be coming, sad to say. They go to Clermont rodeo this week, so with any luck they might come and see us, though it is a long trip out of their way just for a visit. Don has been doing very well so far this season.

Unfortunately one can't muster every day of every week, but I could ride for days now and not become tired. It's the coming back at nightfall and having to race round getting tea ready that gets me down. I know now what a hard-working man must feel when he returns tired and hungry from his office and has to start cooking for himself.

20 September

Yes, I can tell you what I would like for Christmas: a tin of good shortbread and a Scottish calendar with a picture for each month. Thank you.

This week the bush is bursting into bloom. The brigalow is covered with mimosa-like fluffy balls, alive with English bees, the sandalwood has pretty white flowers and a sweet scent; then there is a bright yellow broom-like shrub, a purple sort of bush that grows in poor country (myrtle, perhaps), and lots of small weeds in bloom, including a pale blue gentian-shaped thing flowering in the middle of the road. In the garden we have one single sunflower, but the beans are rioting all over and around the gas cylinder and its gadgets, so heaven knows what will happen when the cylinder needs changing.

Last Wednesday we went in to Alpha to bring Mrs Carruthers home, as she was fed up with the flies, the heat, the gossiping Alpha ladies, the food and especially the doctor. He thought she was in for a 'cure' for her arthritis, which she has had for thirty years. When Merv asked him if his mother could come home he said, 'Her arthritis is not better yet, so I'd like her to stay a few more days'. I ask you!

Next day Merv and I went out to the tractor and took the meat and bread out to the three fencers, all cheerful fellows. We found their fridge plonked down in the dust at their camp with a huge gidyea post holding the door shut. They are getting on well with the fence, with about six miles of posts up and bored. While Merv worked on the tractor I cleaned the other trough at Clara's Dam, and while running in the clean water I found a 4-inch freshwater mussel blocking the float pipe. There was a pelican, of all things, floating on the dam. He must live on mussels and eels.

Do you know, Merv hasn't seen his western boundary yet, and doesn't even know if there's a fence there at the northern corner. He wondered why there were so many brumbies on the place, and the fencers asked if they could be coming over the ridge near the boundary. Merv said he hadn't a clue as he'd never been out there. It's incredible really, but it may give you an idea of how wild and rough the country is out here. On our way back the men asked us to have a look in their camp oven and check that the stew was all right. It looked horrible, but smelled okay: cabbage, meat and onions floating around in a lot of grease. They put it on in the morning before setting off to work, leaving half a tree smouldering under it. It simmers on and off all day, depending on the breeze, and is more or less ready for them when they come back at sundown.

Merv found a snake in the shed yesterday, about four feet long. There seems to be a plague of mice after the feedbags, so the snake must have been living on them; maybe we shouldn't have killed it.

The Boss sold seven bullocks yesterday for $122 each to a buyer who came round, then kicked himself for not taking them to the Clermont sale today, where they would very likely have fetched $140 or $160 each. The demand for fat bullocks is tremendous just now.

A new dogger has come to take old Paddy's place; his name's Joe. He has a newish Landrover, a dog called Boxy, and an apparent wealth of knowledge about the bush and the tricks some people get up to. There's a lot of cattle-duffing going on, presumably because it's an easy way of making money as it is difficult to prove. Usually the cleanskins are taken at night and branded by daylight, so who's to prove they don't belong to the rustlers?

29 *September*

The weather is starting to heat up at long last, reaching about 90°F at 2 p.m. and dropping to about 80° at bedtime, which seems cool. It gets cold at dawn but warms up again quickly. The curlews have been calling in the night, sounding like squeaky swings going to and fro. It's an eerie noise, but I like it in an odd sort of way, just as I like the sound of howling dingoes, even though it makes my flesh creep. Incidentally Joe, the new dogger, is apparently an outstanding fellow. He could be manager of the best station in Queensland, but has a mentally defective wife, which is why he has taken up dogging. He's very different from Paddy – who died, by the way – and instead of the inevitable rum, he goes around with a bottle of Dettol.

Last Wednesday Merv, Drew and I went out on the tractor. It ran out of fuel, so I had to go back in the 'rover with Drew to fetch some more. Drew didn't want to come with me and was most upset, calling me all the names under the sun on the way home; then he bumped his head on the dashboard and began to scream, mostly with temper, so I had quite a time driving, slowing down for creeks and gullies, avoiding cattle and controlling the boy with one hand. At the homestead I picked up some fuel, thankfully put Drew to bed and went back to the tractor. On the way home again I cleaned a couple of troughs and filled them with fresh water, so now all the troughs on the place are clean.

Sunday was anything but like a day off, but it was quite exciting. Before lunch the three of us went looking for the lame cow to slaughter that night, and as I was lookout, standing on the back of the 'rover , I saw the beast lying down in the grass. When Drew was asleep that afternoon we crept out and found her in almost the same place. Merv shot her and cut her throat, attached her to the Landrover and dragged her to a tree, where we strung the carcass up on a branch with the block and tackle. Merv then gave me a knife and I skinned one side quite successfully – the poor thing was in calf, too – and then she was split down the backbone with a hand-saw, at which point we were both sweating, and the mozzies and blowflies had arrived in their thousands. The four quarters were laid on a bed of branches in the 'rover and we were about to set off for home when Merv noticed – would you believe – a flat tyre. It was almost dark by then. I jacked the vehicle up while Merv took the wheel off, only to discover that the spare was flat as well, with a stake hole in it. So, with headlights on and the insects multiplying by the million, he started patching the inner tube. While he was thus engaged some cattle came across the cow's spilt blood. They bel-lowed their heads off, their mates came hurrying from all directions and the lot of them snorted and groaned and pawed up the dust, coming closer and closer in the dark. It was quite terrifying, es-pecially as they were hidden from sight. They were certainly stirred up and restless. I was ready to jump into the 'rover at any moment. but at last the wheel was ready, the engine belted out its comforting hum and off we went. Naturally we couldn't find the road, however, as the bush looks completely different at night in headlights – just a sea of yellow grass and silvery trees all around. Finally we found the track, much to our relief, and by 8.15 we were home. We then had to hang the quarters up in the shed, which was heavy work. Still, it was a good day off. I'm an apprentice slaughterman now, of all things! Yesterday we were up at 5.15 to butcher the meat – I did the corning and boning – and we had some delicious fillet steak for breakfast.

Today while Merv was out on the tractor and we in the house were resting, a boy from Dalgety's agency came out from town in answer to the request for a stockman to help with the next muster. I took him out in the 'rover to find Merv, and he got the job. His name's Trevor, he looks about 15 but says he's 18, and he seems rather a silent type.

4 October

During tea on Thursday a violent wind storm blew up and sent papers and mats flying through the house. Then there was a dreadful, eerie banging and thumping noise that got nearer and nearer while we all sat in frozen silence wondering what the hell it was. There was a final thud, and then it stopped. It was an empty 1000-gallon tank that had rolled over from the shed and come to a halt against the fence. There was a lot of lightning in the south but no rain for ages. Finally it did come; it was only 63 points, but has made all the difference to the growth already.

On Friday Merv and I set off for Forrester station in the truck to pick up a few of Wendouree's cattle that were in their yards. The road was wet, and we succeeded in bogging the truck once in slimy black mud, which in turn cut up the road a bit when we extricated her. There were nine head to be loaded, including a couple of half-Brahmans that glared ominously at us and waved their horns. However, they were successfully chased up the ramp and we went home with no further trouble. Back at these yards we unloaded them, and branded and earmarked the two calves.

Trevor, the new boy, is quite a willing lad and seems pleasant enough, but like most of the other males I've met so far, he's been trying to get off with me – with no success, I may add! He's only 17 and not exactly my type. On Friday night we heard him shooting away at something down at the cottage, and next day we learnt that it was the dingo trapper's delightful dog Boxy, who had walked 16 miles from the fencers' camp where he's been staying while Joe is in town. Just as well Trevor missed in the dark; he thought it was a dingo about to jump into bed with him! Boxy stayed with us for three days, and then yesterday went walkabout again, looking for his master, no doubt.

I shot two goannas near the chook yard; I used only one shot on each. One was over six feet long. Later I took Boxy to get the cows and saw a young jabiru stork walking slowly along the bank. I stalked (ouch!) him and took a photo, and then Boxy couldn't contain himself any longer and leapt at the bird, chasing him into the creek, where he fell forward onto his big beak. I helped him out onto the grass, but he just lay there looking very sick. Next day he'd died without moving from the spot. One black chook and a laying hen died yesterday too, for no apparent reason except that they were very thin.

Yesterday was spent in the bush; it was a great relief to get out of the house for a change. Merv, Trevor and I went along the southern boundary and found about six breaks in it from brumbies running at night. We fixed these, and went on to an old fence-line to recover the barb from it. I cut the wire off each old post, Trevor straightened it and Merv rolled it. Meanwhile masses of brumbies came up to have a look: lovely horses, all shining and healthy and wild, mostly chestnuts and bays, with a few blacks. After lunch a mob of nine came up to the fence and Merv decided to shoot them. He killed the lead mare first, and the stallion came right back to her bringing the others, so that they all died immediately except one bay mare, who had two bullets right in her chest yet still kept going. Later, when all the others were just heaps of meat, this poor mare was still standing. I begged Merv to finish her off, which he did. Such a wicked waste of horses. I was nearly sick, but realise it has to be done. We went on with the wire, eventually recovering one and a half miles of it, but my mind wasn't on the job at all after the massacre of the brumbies.

We have had beans every night for about a month, but only had two feeds of peas before they wilted and died; also some nice beetroot and lots of carrots. The sunflower plants are over ten feet tall and look in at the kitchen windows, much to Mrs Carruthers's delight. She is much better at present, so we're not in so much of a hurry to go south. However, Uncle is still coming up as soon as he can, and we'll get the weaning muster done and the bullocks too, as Merv will have to sell them.

I punctured my toe on a bone the other day, which makes me think a course of anti-tetanus injections wouldn't be a bad idea one of these days. One is always getting cut and scratched, and can't avoid getting dirt into the wounds. The toe is okay now, but it swelled up a bit and was quite sore at the time.

13 October

My birthday began at 5 a.m. yesterday, when Uncle stuck his head round the door and said, 'Come on, kid, it's late!'

Thus started the big muster. Trevor saddled Twinkle for me and brought her into the garden while I washed up, fed the chooks and swept the floor. In the end I was ready first, so rode down to the yards to find Flicka bucking herself off under her saddle, the stirrups

Kabunga West, Wandoan, home of Jessie and Owen Perrett.

Dinner camp on the Dawson River: Dick Perrett and his nephew Bruce.

Wendouree homestead.

The view from my room at Wendouree, looking out to the front gate, landrover and some of Toby's horses in the background.

Merv Carruthers and his tractor.

Main Street, Alpha.

Twinkle, Flicka and me.　　　　　　*Doug and Don at Home Hill rodeo.*

Picnic at the Rocky Crossing: Merv, Terry, Uncle, Drew and Jean Swinbourne.

Milking.

Boxy, Joe's Australian blue cattle dog.

Some of the cows old Uncle and I brought to water at the Rocky Crossing.

Blowout on the black-soil plains on the way to Toobrack, near Longreach, to buy bulls.

Sandy Creek, which is usually dry, after three or four days of rain.

Bogged. Standing beside the vehicle is Uncle.

Greentrees waterhole is a favourite watering place for brumbies and budgerigars.

Kate Doonan's grave, 1885. She must have been one of the first settlers living in the area of Wendouree. Her husband managed an inn on the old mail run.

Young cattle Droughtmasters.

Castration in progress, Wendouree yards.

flying about and thumping her in the ribs, which made her buck even more. She was only protesting about being saddled after such a long holiday. By seven we were all aboard, leaving Merv looking quite out of place in his beloved Landrover. However, he manages to do quite a bit of nifty cattlework from behind the wheel, more than he would do from the back of a horse.

Joe the dogger is coming to work for Merv as soon as he can buy a house in Alpha for his family. He gave us a couple of days this week meanwhile, and rode Flicka. He's about 50, knows a great deal about bush life and keeps me enthralled with tales of cattle, horses and dogs – though I'm not sure they're all absolutely true.

The four of us, Trevor, Uncle, Joe and I, picked up twenty head just outside the yards, which were put straight in without further ado, and set off for more. I found ten on my own later with no trouble, for Twinkle is just the nicest horse I've ever had the pleasure of meeting and is no problem to control. I can ride her with hands in pockets and one leg hitched over the kneepad down her withers, almost side-saddle fashion, but unfortunately she doesn't look where she's going and tends to fall over stumps and things unless I guide her everywhere. At one time, when we were crossing the dry creek in pursuit of a weaner (a heifer of course, they're nothing but trouble) that ran along the far side of the bank, I very nearly came off. The heifer suddenly stopped and changed direction under Twinkle's nose, and she turned in the same instant, hard on the little beast's heels, leaving me almost continuing in the original direction.

Later we were charged by a big, red, horny scrub-bull, who shook his head at us then came full pelt. Twinkle took matters into her own hands and raced straight at him. I felt like a not very brave lancer without a lance, but fortunately the bull turned first; very fortunately, because he was almost as tall as the mare and twenty times heavier. Her black coat was covered in sweat and streaked with dust by then, but she was still keen to chase anything that moved. By midday it was getting hot, and we had about 400 in front of us, so we started to take them home.

All at once I found myself alone with the mob, as Uncle had raced Nugget off to the left, Trevor disappeared to the right and Joe was somewhere back in the creek, all of them supposedly picking up more cattle. Nothing disastrous happened in the ten minutes we were alone, however, as the heat kept them all fairly apathetic; in fact it was difficult to keep them moving. Uncle raced around cracking his stockwhip at everything (which Twinkle hates; she flinches and tosses her head at each report), and the little calves dropped back to the rear of the mob, panting and foaming pathetically. By two o'clock we at last reached the yards and pushed the cattle in before going to the house for a cold lunch. Then we returned to the yards to draft the cattle, and let the horses go after washing their backs and giving them a small feed of grain.

That night, after getting tea and flying around, we had a 'party'. I made myself a cake and iced it, and after we'd washed up Merv opened a bottle of sparkling white wine and everyone sang 'Happy Birthday', which was a trifle embarrassing but very nice of them. Mrs Carruthers, bless her, had ordered from a mail-order firm a pair of very pretty, pale turquoise shortie pyjamas with pink rosebuds on them, and she presented them to me with the sweetest smile on her careworn old face. Merv says he's giving me a new pair of riding boots, but they haven't come yet. People are very kind.

Uncle arrived back last Thursday, and on Sunday we went out in the 'rover to catch the horses, as there was no night-horse to do the job. Uncle could only catch Twinkle in the open, so he put me up on her bareback to lead the others home. All went well until we reached the creek, when Billycan belted off, followed by the others, and it was all I could do to hold Twinkle back. She shook and shivered and fretted until Uncle came and held her head; then she trod on his toe and made his finger bleed. (That sounds Irish, somehow!) Merv set off after the horses in the Landrover, and drove them right up to the yards, while Uncle told me to go and do the actual yarding-up. Twinkle cantered the best part of two miles before successfully pushing the others into the yards, so I was quite chuffed with life in general, though suffering from 'rawhide', as she has the most prominent withers and backbone you ever saw. 'You've got an extra saddle now, Merv – this girl don't need one!', said Uncle. Praise indeed from a one-time champion jockey of Queensland.

Remember I told you I had written to the Hassalls' married daughter at Meadowbank, Mt Garnet, to see if I could go up there after Christmas? They're friends of Don, who was working here when I first arrived. I had a letter from her in which she asked where I was spending Christmas and whether I would like to go there and spend it with her and her family. They might need help in January, but even if they didn't, would I like to come just for fun? She – Glen Rankine – sounds so nice, and I may well take them up on their offer, though I'm still going down to Jean's when Merv goes to Sydney as soon as this muster is over. Joe says the further north you go, the more friendly the people are. He also says January is the worst month of all to go up there; one gets cut off by floods for months, if you stroke your horse's neck your hand comes away covered in blood from the mozzy bites and it's as hot as hell until

June. To me it sounds marvellous, and I look forward to the challenge.

The day after my birthday we were up at five to start the yard work, but Uncle was really drunk, for the first time to my knowledge – apparently from the one glass of wine he had the night before, topped up with a rum or two when he rose that morning. He was going like a greyhound all day, jumping on calves' backs and shouting 'Horsey, horsey!', and brandishing the branding irons at all and sundry. Everyone kept well out of reach until his enthusiasm evaporated.

I did some of the earmarking of the forty-one calves, washed the cutting knife between each bull calf and kept score of how many steers and heifers there were. They were weaned as well (too many traumas all at the same time, one would have thought) and Merv and I took the weaner heifers out to Bullock Head Paddock in the truck – a distance of 14 miles. My back was practically raw from the rough going. Twice my head touched the roof, and at one point six weaners fell down and had to be prodded to their feet again, otherwise they would have been trampled to death by the others.

It's Friday today, and my writing is becoming worse and worse. The temperature inside the house is 94°F, there's not a breath of wind and storm clouds are gathering. I've had a hectic morning taking the weaner steers out of the yards to water all alone. Twinkle and Jim were saddled and I waited with them at the yards for an hour, expecting Uncle to arrive. He'd gone off with Merv, however, so I finally said, 'Dammit, do it yourself', opened the gate and mounted Twinkle. Ignorance is bliss . . .

There were about eighty head of unhappy calves all longing for their mums, and once they got through the gate they charged off to the right in the direction the cows had gone. I had to race through the trees and try and turn the leaders towards the creek, but they were a determined lot and split up in all directions. Somehow I got most of them down to the water, where they drank thirstily, but by that time Twinkle, to my shame, had a girth gall, so I walked her back and mounted Jim. Meanwhile one blind steer was still in the big yard. I couldn't get him out, as he had no idea where to go and kept crashing into the rails. Once he almost tripped Jim up by going straight under his belly, so I left him and tried to round up the six or seven that hadn't gone with the mob but had spread well out. They just would *not* go together. One or other of the little perishers would keep stopping, bawling mournfully and then setting off back the way it had come. By the time we got them all down to have a drink, Jim and I were literally dripping with sweat. That's the last time I attempt to move weaners single-handed.

I saw a huge brown snake disappearing down a hole yesterday, and another had been squashed on the road. They are everywhere just now with the warm weather. Oh yes, another little snippet of useless information: there are six baby turkeys now to swell the ranks in the chook yard.

21 October

Last Sunday was once again anything but a day off. We were up at five to a frantic morning of doing the washing, changing all the beds, feeding the horses, baking cakes and preparing lunch. After that, instead of resting, Uncle, Merv and I set out to find a heifer with an umbilical hernia to kill for meat. We combed Doolan's Paddock and found two neighbouring cows that were too poor to kill – it's a fact you have to go next door to taste your own beef – so we went on into Charlemont Paddock, where just on dark we found the roan heifer. On the way I spotted a dingo, which Merv shot dead at 300 yards.

As we shot the heifer and skinned her, big storm clouds gathered over the livid sunset and Merv remembered that the petrol gauge was low. Suddenly a great gale sprang up. We raced to get the meat quartered and loaded, but the rain came down in sheets, hastening the dusk and darkness followed quickly. As we piled into the Landrover to go home, Merv said, 'Bad news – no lights!'. The battery was flat. So, with 12 miles to go and a very rough grassy plain to get out of first, full of stumps, anthills and gullies, we were in a fix. Uncle and I got out and tried to guide the 'rover over or around the rough stuff, while being drenched with the freezing rain, and slipping and falling about in the mud. Eventually we navigated the vehicle onto the road, though we couldn't see at all except during the lightning flashes. We got in to shelter, but steamed up the cab like a couple of wet spaniels after a day's shooting, so got back out again and walked to the Gidyea Crossing, hardly able to stand up for the raging wind. Shortly after crossing the muddy creek the engine cut out and wouldn't start again for want of fuel, so we hung the meat in a tree and walked the rest of the way home: 9 miles, which took two and a half hours in the pitch dark. When at last we got home I was literally dizzy with hunger and cold. Mrs Carruthers was frantic with worry by the time we turned up, at 10.30, like a trio of drowned rats. I thought Uncle would get pneumonia, but he's a tough old stick and was none the worse. I fell into bed, but poor old Merv had to get up at 4 a.m. and take the truck out, with Trevor, to collect the meat and the dead dingo and bring them home before daylight. The butchering had to be done before sun-up, when the blowflies would arrive.

On Tuesday we all went mustering, except Merv, who had strained his back. We picked up a small mob, and Joe told me to *lead* them down to the crossing along a cattle-pad. I was dubious but rode ahead, and much to my surprise all the cows strung out behind and followed very well. Joe introduced me to wild 'black currants' and the gum off a wild orange, which was rather tasteless. He says gidyea gum is best when you're hungry.

Twinkle became very lazy with the heat, stopping hopefully under every shady tree when we were taking the cattle home. We had one race after a red Brahman mickey and got him back to the mob successfully, and lots of fights broke out among the numerous bulls, which were soon quelled by cracking whips.

There were 350 head in the herd that day, so the next day was spent in the yards. One heifer got up from the branding cradle, saw me first and charged. I bravely raced behind the nearest man, and she stopped at his feet. Later we took the heifers out in the truck – three trips all told, which was very tiring – and I came back to fetch the milkers, feed all the various animals and bath Drew, which was the hottest job of them all, especially when he doesn't want to be washed.

Yesterday we mustered again, and it was another mad rush to make breakfast, hang the corned beef, start the engine, feed the chooks and turkeys, get Mrs Carruthers up and half dressed, and wash up – during which time Drew was dressed by his father and Trevor got in the horses. The boss took us all down to Doolan's Bore in the truck to save the horses an unnecessary 6 miles, and we rode on from there. I found a black cocky's feather with the bright red stripe in it and stuck it in Twinkle's headband, which made her look like an Indian warhorse. At the creek we split up, and I went with Joe through the gidyea and sandalwood scrub – thick-timbered country with washouts all over the ground and lots of hidden holes. Joe is a pleasant companion to ride with, a real old-fashioned type of bushman who rides along playing tunes on a leaf and talking quietly, watching all the time for tracks and signs of cattle. We got about thirty, which we took to the dam. There we met Uncle and Trevor, who had only found one bull; he promptly had a fight with the one we had. The sky clouded over on the way home, and we were well and truly bushed without the sun to steer by, ending up by the fence we had started from. The cows with very young calves were dropped off, as they were flagging, and we turned round and tried again, naturally getting back a bit late though we managed to yard up before dark. Poor old Uncle had an upset tum and was sick for the last 2 miles. As soon as we got home, he went off to bed, leaving me to deal with Billycan. It was another half hour before I was able to drink a welcome pint of cold water.

29 October

Well, the hot weather has come at last, and everyone seems to be knocked up by it except me. No doubt I'll collapse when the others recover, but it *is* nice to sweat again. I only need a sheet at night, until dawn, when a blanket is welcome. However, one might as well get up at that time anyway, because the horse in the night-paddock whinnies non-stop to be taken out, the milkers bellow for their feed, the cockies and parrots screech their heads off coming in to water and of course there are two dozen roosters all greeting the sun. What's the point of staying in bed with that infernal racket going on?

We mustered on Monday, after the men had tried to run the horses in by using the Landrover, and failed. Twinkle was in the lead, and took them on a long detour before running into the milkers' yard. She waited until someone came to shut the gate, but raced out to freedom along with the others at the last moment. I caught her later with a sly bit of toast, and ran the others in. Joe bought three new horses for Merv last week and rode one of them, a real film star, big-boned, with a fabulous creamy chestnut coat and large liquid eyes. He's called Goldie, and he walked out as good as gold (!) until, an hour later, he put his head down, bucked violently on top of a flat anthill and fell down, throwing Joe clear. Joe was okay, but Goldie got a black eye. Twinkle was in good form and hard to hold at times, though she slowed down in the heat of the afternoon. We picked up a whole lot of scrubbers, which were very difficult to drive as they kept breaking back to the scrub.

I had a long conversation with a big polled bull the other day; he started it by bellowing at me, so I copied him while he got more and more irate until he started pawing the dust with his head lowered, at which point I beat a hasty retreat before he actually charged.

On Thursday we went into Jericho for a cattle sale, just to see how the prices were going. I drank a bottle of lemonade – the first one I've had in this country – and it *was* good. Of course such things are luxuries in the outback, and it did make a lovely change from muddy, lukewarm creek water.

5 *November*

Last Monday I went out with Uncle to pick up a few more cows in Doolan's. He used Twinkle to run in the other horses and had an awful time with her; the poor mare was sweating all over; possibly I've spoilt her. The wretched new horse Goldie caused most of the trouble, racing all over the place and always in the wrong direction. It was nine o'clock and the sun was well up by the time we set off, but we mustered over eighty head, surprisingly – there must be a break in the fence somewhere – and took them up the creek, losing about a dozen in the young gum suckers. This is an unforgivable crime for anyone who considers himself a good stockman, but they outwitted us quite easily. We held the mob on a waterhole during lunch, which Twinkle shared with me nose to nose, eating the bread from around the sandwiches as well as her own ration of dry toast. I just drop the reins now, and like all good cow ponies she just hangs around. She investigated my quart pot but obviously disapproved of the hot black tea within, and snorted loudly, making me spill half of it.

We had trouble getting the cattle to move afterwards because of the heat, and I was dying of thirst from the dust and yelling and cursing until I found some blackcurrants, which helped considerably. The sun burnt my back through my cotton shirt as if I wasn't wearing anything at all. It was 5.30 before the mob was yarded, and then I had all the animals to feed – a nice job, but time-consuming at the end of the day.

Next day after the yard work we took the cows and young calves into Charlemont, with more curses, as the weaned cows kept turning back in a most determined fashion. On the way home Twinkle sneezed and drove a stake up her nose – ow! She dripped blood for a while but is okay now. She's very absentminded and doesn't look where she's going, so she's always in trouble.

Suddenly we came face to face with a huge red roan stallion with a tangled black mane – the same one as before. It was thrilling to watch him poised motionless for a second before he started dancing and rearing before us. Behind him were three black mares and two bays, all lovely creatures. He wanted Twinkle to join them, even though she wasn't in season. I was just going to get my faithful Instamatic camera out of the saddlebag and take a photograph of him when Uncle came charging along and chased him and the others off. It was a pity, in a way, but it was probably a good thing, because the stallion looked as though he meant business, and things might have got nasty.

I have been riding Twinkle bareback lately as she is now resident night-horse and it's convenient to fetch the cows on her. I use a blanket folded in four on her back and mount from an anthill and, do you know, she stopped at one just the right height yesterday! A most considerate beast.

No one else can catch her now, out in the paddock. She's a crafty old thing; she keeps just out of Joe's reach, which infuriates him, and Merv has the same trouble. She walks up to me, though, sticks her pretty head in the bridle and waits. She's very lively when we're after horses, and it's difficult to stay on her, but she's quite different with cattle; she just plods along until a burst of speed is called for.

On Thursday I slept badly and awoke with the sound of a car arriving at midnight, which stopped long enough to drop something off before driving away. We found out later that it was the fencers, who were passing through and had left two of the sweetest kittens for us, as they had promised some time ago. One is tortoiseshell and white, the other pure white, and both have grey eyes. Drew got hold of them, unfortunately, and has succeeded in dislocating the leg of the white one and half drowning the other in hot water. It's time that little boy and I had a few words on how to treat animals.

Rodney Perrett wrote last week. He sounds as mad as ever, is working hard on his new block and is up to his ears in debt, but seems quite happy. He says he has all the cattle and horses he needs; the place is running with brumbies and ancient scrubbers.

12 November

On Sunday I dragged Merv off after lunch to clean Greentrees' troughs, as stock will be going in there this week. While sluicing one trough a dark green and orange centipede climbed up my leg. It fell off again almost immediately when I shied at least five feet sideways with fright.

At dawn on Monday it started to rain, pouring down all day and developing into pretty violent storms that night, with about two inches falling near the homestead, setting the creeks running again. By Wednesday it was clear, but the ground was heavy underfoot. However, we went mustering bullocks, taking the horses in the truck to Doolan's Bore to save them 12 extra miles.

Merv rode Jim – very reluctantly – and got a sore bottom, knees and thighs, and pins and needles everywhere else. He missed his Landrover sadly. At one time I had to race Twinkle to head off a stampeding mob of steers. My heart was in my mouth at times, as the ground was hollow underneath in places and full of 'melon holes', but she didn't put a foot wrong, bless her.

Next day we went over the same country and found fifty-three cattle in the scrub. I found a pad with fresh tracks on it so just followed it into the gidyea, where there are lots of boggy gilgai (shallow waterholes after rain). I wasn't taking much notice of where I was going and suddenly found myself face to face with five steers, all watching me. They raced off with tails up, Twinkle hard on their heels, and after much ducking and winding through the trees and jumping fallen branches we emerged onto a fair-sized open flat that I had never seen before. Having no idea where I was, I tried to turn the fleeing steers to where I thought the fence might be, but they all split up. Twinkle got them together again with some nifty footwork and headed them in one direction, and to my relief, some time later, we hit the road. I settled the tired cattle under a shady tree and waited until they were resting before setting off back to look for Merv. Twinkle went most of the way like a bloodhound, her nose on the ground, belting along at a fast trot and whinnying every so often as if calling Jim. She knew who we were looking for, and we found him, only to find that *he* was lost. The horses knew which way home lay, though, and took us back to the five steers by the roadside. Archie and Joe had already taken their mob home, so we had a quick lunch and went on with our five – what a big herd for two riders to control!

During the night the bullocks flattened a section of the inner yards, breaking the posts right out so that there were cattle in about four different yards. They were milling about uneasily when we let them out in the morning, but soon settled down and allowed themselves to be driven along without fuss. Joe, Uncle and I had to take them way out to Greentrees, a distance of eleven miles or so, and when they got there they all drank from my nice clean trough, then put their heads down and ate all the herbage round about – the new pigweed and sweet grasses that have sprung up since the rain. After lunch at the creek we rode back to Sandy Creek Bore, where Merv was waiting in the truck to take us home. Sugar, who I was riding in order to give Twinkle a rest, had never been in a truck before, however, and refused point blank to try it. I pulled, Joe and Merv placed her dithering feet on the ramp and Uncle got behind with a big stick, but all to no avail. It took over half an hour to load her, and by then she was a sweating wreck. So were we!

Unfortunately Boxy, Joe's blue cattle dog, has killed the white kitten. Drew dropped the other one in the calf's bucket of milk and was about to dig its eyes out with a chisel when I rescued it.

19 November

To answer your questions first:

1 The new place of work is called Meadowbank, a cattle station running Droughtmasters (a Brahman–British red cross). The nearest town is Mt Garnet, which is south-west of Cairns. The people, Mr and Mrs Hassall, have a married daughter, Glen Rankine, and two small grandchildren called Tammy and David. I am to share a room with Sharon Kruckow, the other jillaroo, who has been there a year already and is staying for a further year.

2 I haven't been paid yet for the simple reason that I haven't needed any money, but on leaving I will get about $500, which isn't bad. The pay at Meadowbank is $14 per week for the first two months, then $16 a week, working only five and a half days each week, with strict time off and a rest after dinner most days.

I am very glad to have been able to help this household these past seven months and get things moving again for them. Now Joe is here the cattle work is streets ahead already, so I reckon Merv is on the way up at last. He deserves to succeed.

I have just discovered there are koala bears up here. Not that any have been seen, just their scratches on certain gum trees, but at least they are about. I'd love to see one in the wild.

On Friday night Merv and I took Uncle in to catch the Midlander home while my two cases went on the goods train. They will take a week to get to Ravenshoe, the nearest railway town to Mt Garnet, so I just hope they are still recognisable by then. At last I know when we are all going south: Tuesday, 29 November – if it doesn't rain in the meantime – to Toowoomba, which is over 600 miles away.

25 November

It is Friday after lunch and I'm sitting on my bed in next to nothing, the temperature just on 100°F and the air very dry and still. The cicadas are the only creatures bothering to make a din, otherwise it's all nice and peaceful. I like the shrilling of cicadas, for although it's a monotonous, penetrating noise that would do nothing for a headache or hangover, it takes me back to the jungle roads in Ceylon.

All this week I have been cleaning windows and cupboards and doing everyone's packing but my own final bits and pieces. On Wednesday we went into town, where I bought a packet of prunes and dried peaches to chew on; one certainly misses the fresh fruit out here. When out mustering in the dusty heat and flies the other day, I suddenly pictured a dish full of nectarines and greengages and nearly fell off Twinkle. In a year or two, though, the young citrus trees we planted earlier should start bearing fruit, and life will be that much more pleasant for whoever takes my place. We've been killing off the roosters, so have had chook for lunch and tea for the past week. Any day now I'll start crowing to greet the dawn.

The horses came up to the house yesterday – the first time they've been over the creek since the muster – so I took some feed out to Twinkle. She saw me coming, put her ears back and headed off for a few steps then stopped. I could see her thinking, 'Ah! She hasn't got a bridle hidden away anywhere, so perhaps it's safe . . .'. Round she turned and trotted right up to see what I had for her in the way of goodies. I guess I've had my last ride on her, as there won't be time this weekend, and Merv hopes to be off by 5 a.m. on Tuesday. Joe will be here to take care of things. He still brings me back some new little thing every time he goes out, like a tiny scorpion in a matchbox, a lump of gidyea gum, an interesting flower or a plain turkey's egg. I've learnt a lot about the bush from him.

Holidays: 314 Queen Street, Maryborough

2 December

Here I am, safe and sound with the Swinbournes, and very glad to be here. Let me start at the beginning of the week. Sunday there was a huge wash to do, with the consequent ironing; then on Monday I was flat out from 5.30 a.m. feeding the chooks, turkeys, cats and folks, helping Mrs Carruthers pack and sorting Drew's clothes and toys and packing them. I finished the ironing, washed Mrs Carruthers's hair, swept and mopped the floors, collected the eggs in an outside temperature of 100°F, burnt the rubbish, washed some clothes, finished Mrs Carruthers's packing, *made a cake*, prepared tea, watered the tomatoes, fed the chooks again, put Drew in the sink and washed him and his clothes, cooked a meal and served it, washed the dishes, sorted eggs and tomatoes to take with us, cleaned the fridge . . . and fell into bed at 10.30. We had agreed to get up at 4.30 on Tuesday, but my alarm was the only one to go off so I rose and finished packing before waking Mrs Carruthers. I dressed her, then Drew, tidied their rooms, stripped all the beds, fed the cats, put cold water and milk into billies to take on the trip, washed up, closed all the windows and we were ready to go. Said a sad goodbye to Joe and the cats and set off, seeing no sign of my dear black mare in the home paddock. Perhaps it was just as well. The heavily laden Mercedes bore us as far as the seventh gate to Alpha, when Merv remembered he had forgotten his suits, so back we went, losing an hour.

The journey was hot and dusty, but we went through some really exciting country. Well, to me it was exciting, being full of gum trees and cattle, dry creeks and lonely homesteads, with the occasional kangaroo or wallaby bounding across the road. We passed through Tambo, Augathella, Muckadilla, Roma and Yuleba (the nearest town to Potter's Flat and Mrs Golden) and at 7.30 eventually hit the big town of Toowoomba. We spent the night in a luxurious motel, where I had a single room complete with TV, radio, adjoining shower, basin and loo, an electric water jug and packets of coffee, milk, sugar and tea – but the brackish water tasted of shells, no matter how one tried to disguise it. I had supper with the others in their room, and was introduced to two girls who were going down to Brisbane in the morning and would take me with them in their car. They weren't my type of female, being too obsessed by clothes and hairdos for us to have much in common, but it did save me having to catch a train.

Next morning a cold and depressing, dank fog lay all around. I dressed and had breakfast with the others – pawpaw, bacon and eggs and milk – washed Mrs Carruthers as she couldn't get her wheel-chair into the shower, fed Drew, got everything straight for them and saw them off. It was really sad leaving them; they seem such an unhappy family below the cheerful surface. I do hope Merv goes from strength to strength, that Drew settles down properly and starts liking the bush, and that Mrs Carruthers suffers less from her awful arthritis in the cooler climate of Sydney. Just before they left Merv gave me a cheque for $477, plus $20 'to buy a pair of spurs with', saying that if he could have afforded it he would have paid me double. He really is decent.

So, I was on my own again. The girls left at nine, grumpy and still half-asleep, but they cheered up when a sergeant from the RAAF hitched a lift with them. He was an exceedingly dull bloke, I thought, but then I wasn't in the mood to make polite conversation, for in my mind I was back at Wendouree, riding Twinkle through the gidyea scrub and sharing a sandwich with her.

The girls were hair-raising drivers – they took it in turns – and after a near miss in Brisbane's biggest and busiest street, I got out where I could, thanked them quickly as they roared off, and went to the station to dump my bag. I then rang Mrs Sinclair at Wavell Heights, but there was no answer. I wandered round some of the bigger shops and bought a few odds and ends, but the crowds! I became quite disorientated. Having sent Jean Swinbourne a wire saying I was arriving on the earliest train, I had a milkshake and took myself to see Marlon Brando in *The Chase*, my first film since March. That filled in the time until five, when I rang Mrs Sinclair again and had a chat. She was going out to a party that night, though, so I wasn't able to see her. Instead I treated myself to fried rice, chicken and almonds and bamboo shoots and sweet-and-sour fish at the Lotus House. It was a bit greasy, but a nice change from steak and mutton. (Hey, listen to me!) Finally I wandered up to the station with a bag of cherries under my arm, feeling dead tired, and climbed aboard the Rockhampton mail train, which left at seven-thirty.

The train pulled into Maryborough at 2 a.m., a ghastly hour; but Jean and her husband Rod and young Terry all met me and were so kind and cheerful, giving me a mug of hot chocolate and putting me to bed, where I slept for six hours and woke feeling much better.

This is a lovely house; the rooms are built above the garage and laundry, all bright and freshly painted, and there's a verandah-style balcony all round. My room looks straight over the back garden down to the Mary River, where the Swinbourne boat *Sue* (no relation) is moored.

I had a marvellously lazy day yesterday, just sitting around watching Richard Attenborough on TV, eating salad for lunch and sleeping for three hours in the heat of the afternoon. Later I walked down to the river and onto the little wooden jetty and looked at the swirling muddy water, which is incredibly deep at high tide, although the sea is over twenty miles away. At six a couple came in for drinks, which we had on the balcony in the balmy evening air, and I had a long discussion with the husband about cattle. It seems I'm missing the bush already.

It's eight o'clock and I'm still in bed – bliss for a change. Today we are going into town and I must get my boots re-soled; on Saturday there is a pony club gymkhana, Sunday a trip to Bundaberg (where they make the rum), and a dance some time, when I'm to meet Graham Patroney, who has been elected by Jean to be my partner. Then there's talk of a boat trip to Fraser Island, picnics, etc., all of which sounds like a real holiday.

There is a sweet caged budgie here, who is so friendly. He rolls his eyes at you while blethering away nineteen to the dozen, and complains noisily if his cover isn't taken off as soon as he wakes in the mornings. There's also an overweight corgi-cum-fox terrier called Gussie, who is quite a character too in her own way.

It doesn't matter if you write to Alpha, as the post office is sending on letters, though obviously they would get here quicker if sent direct. From Christmas on, would you write to Meadowbank, please? I'm thinking of going up by tourist bus, but Rod is trying to get a crew to take his boat up as far as Mackay after Christmas, so I might sign on as cook and go with them. Or I might travel by car with Jean to Mackay and get the bus to Innisfail. Anyway, no more for now. I really must get out of bed and stop being so lazy.

7 December? – Thursday, anyway

After the gymkhana we raced home for a shower and a quick salad, changed and went to the dancing contest where I was to meet Graham, the chap who had been laid on to show me round. I was prepared to be bored, but in fact it was very good; every kind of dance was performed from the cha-cha to the swing waltz, the hot dog and the Canadian two-step, and Graham was in quite a lot of them. He's a smooth perfectionist as a dancer, but off the floor he's a well-scrubbed, quietly spoken, pleasant soul, very easy to get along with.

Next day we had arranged to have a day out in the boat, and Jean invited Graham along. Rod Swinbourne was up at 5.30 a.m. to prepare for the trip – I gave him a hand to clean out the boat and do a few little jobs – and at length we cast off and motored downriver to the town wharf where the others met us. *En route* I saw a platypus – great excitement. At least we only saw his back and a swirl of muddy water as he dived down, but Rod said it was definitely a duckbill.

The *Sue* is a lovely boat, 35 feet long and home-built (Rod is a marine engineer, which helps), with four bunks, galley, loo – or 'heads' as it's called – stern and aft decks and twin inboard engines. At the entrance to the bay we picked up a young engaged couple and crossed the deep blue choppy sea to Fraser Island, where we anchored and jumped in for a swim in clear green water over white sand. We were burned within minutes. I was surprised not to see any palm trees on the shore, just scrub, which came down to the high-tide line. Shortly afterwards we went on to another place called Woody Island, where we had lunch and iced beer from the *Sue*'s fridge, after which Graham, Terry and I tried to run down some of the herds of wild goats abounding there, but of course got nowhere near them and only succeeded in tiring ourselves out.

All too soon the sun was dropping in the west and it was time to leave. Graham and I steered and navigated the boat right back up the river – in the *dark*, would you believe – following beacons and red lights and avoiding trouble, more by luck than good management. On arrival back we changed into evening clothes and went out to the dancing club's Christmas party, which was quite good, except that we were a bit late and missed some of the dances. However, I learned the polka, rock 'n roll barn dance and the Oxford waltz.

Jean took Terry and me to the dress rehearsal of *Carousel* (local dramatics), where we met the cast. The acting and singing were good, but the story itself was a bit too sad for me. Graham joined us at half-time after his cadet meeting.

I have never been so utterly embarrassed as I was on Tuesday, when Jean took me down to the local sugarmill and left me *all alone* to be shown round by a tall, skinny fellow, whose mates gathered round and grinned and whistled when I had to go up and down steep iron ladders in my green and white shift. The mill itself was quite interesting, but I was glad to get out and bolt back to the car.

Graham called for me later to take me to one of his dancing classes. I was a bit apprehensive about this, but in fact it turned out to be a super evening. They are a nice crowd at the club, all so friendly and kind to an obvious newcomer. I learnt the evening two-step, twilight waltz and part of the hopple popple, which is the best exercise ever invented and not as juvenile as it sounds.

Yesterday Jean and I went for morning tea with Mrs Hyne, the millionaire timber chap's wife, in a huge, lovely white house with big trees and spreading lawns – a pleasant hour spent in the most gracious surroundings. Mrs Hyne has invited me to dinner some time and to look over the sawmill. Hope it won't be a repeat of the sugarmill episode; I shall wear jeans just in case.

I'll be spending Christmas somewhere between here and Mackay, as the sea trip up the Reef is definite. I've booked on a Pioneer Express bus from Mackay to Innisfail for 2 January, hoping to get to Meadowbank the next day.

Thursday, 15 December

Graham took me to see my first drive-in picture show – we saw *She*, a quite abortive film. The second film, *Rhino*, however, was lovely, with lots of wild animals in it. It was most odd sitting in Graham's little Volkswagen with a small box attached to the window for sound. Still, at least we weren't disturbed by people talking and crackling sweetie papers throughout. One odd thing that does occur is that if one car toots its hooter – accidentally or otherwise, usually the former – everyone else joins in, making an instant clamorous chorus for a few seconds.

The following afternoon a crowd of us went round to neighbours who were having a barbecue in honour of the cast of *Carousel*, which proved to be a great party. Graham and I ended up doing an exhibition of the polka and hot dog on the grass in bare feet. Sausages, tomatoes, onions and bread were dished out, masses of cold beer flowed, everybody sang and danced and generally had a good time under the brilliant stars and the party reluctantly broke up in the wee small hours.

Next day, after a necessary lie-in, Jean and I went down to Hervey Bay to swim, but the sea was just a bit cold and too choppy for enjoyable bathing, so we sunbathed instead, explored a bit and ate sandwiches and drank beer.

Rod Swinbourne lived in Malaya for a while and loves curries, so I had to make one yesterday. Despite the fact that I had no chillies, ginger or garlic, it turned out quite well. I used capsicums and red pepper and it was surprisingly good; Rod loved it anyway. He was in Singapore the day it fell, escaping on a boat that was the only one to get through safely to Bombay. It was part of the convoy in which HMS *Perth* was sunk.

Yesterday we visited a Scottish woman and her two nice children. They have asked me to go for a ride on Tuesday, which will be marvellous. I'm to ride the horse that threw the little boy and knocked him out for a couple of days.

Today we went down to the beach again and it was just perfect: a pale green sea and beautifully warm, with no visible jellyfish. We swam a lot and sunbathed and came home for lunch and a sleep. It's a great life.

It still doesn't feel like Christmas, though the radio belts out the occasional carol and jingles a few bells every so often to remind us of the festive season. The sun's so hot though that one cannot imagine hot roast turkey with chestnut stuffing and rich plum pudding with flaming brandy. A leaf of crisp lettuce and some watermelon would be quite enough.

MV Sue, *Barrier Reef*

23 December

I must warn you here and now that this letter will be extremely long because so much has happened since last I wrote. Thank you for your letter, which I received on our return from a boat trip to Bundaberg. It had been sitting in the mailbox since Saturday, and that evening there was the worst hailstorm on record in Maryborough; consequently, your writing was rather hard to decipher as all the ink had run. The storm ripped off tiles, broke windows, brought down trees, made large holes in the golf course, bruised cattle and horses and scared everyone half to death. One local horse had to be destroyed, as it was so cut about the eyes it just went mad. We missed it all, being away, and luckily the Swinbournes' house escaped damage. The neighbours came over afterwards to inspect the place, and found Gussie, the little dog, cowering between two flower-pots shaking like a leaf, and the budgie hiding under his seed box with his wings over his head. The hailstones were the size of cricket balls; it must have been quite terrifying.

On Friday Graham and I went to meet Ian, aged 17 – the second youngest Swinbourne son – at the airport, as Jean and Rod were going to a party. He's in the Navy down near Sydney – a smashing lad and very good company. In fact the whole family is super, and this is just like a second home to me. I gave them your message about visiting you one day and they were delighted. Jean has been longing to go overseas, but has neither the contacts nor the time, and Murray (20), whom I haven't met yet, wants to go some day, so perhaps they will, now that they have someone to stay with in Scotland. Jean is going to drop you a line once things settle down a bit. (Please excuse this awful scrawl. It looks as if a drunken spider with inky feet has been weaving across the page, but the swell seems to be getting stronger.)

It was Saturday morning when we all set off for Bundy in *Sue*. We had a good trip, stopping at Woody Island on the way for a gorgeous swim, lunch and a beer, and reaching our destination by about 5 p.m. That night Graham took me to a drive-in again, then we returned to the boat to sleep. I had one of the forward bunks, while Graham and Ian shared another of the bigger bunks head to toe, which must have been very uncomfortable for them both. Next morning the three of us drove off to Bargara beach for a swim, but there was too much seaweed for my liking. We came back to the boat for lunch, which I prepared: fried parrot fish, sweet corn, pimentos, tomatoes, followed by fresh rockmelon and milk, all of which went down very well. Then we drove back to Maryborough, leaving *Sue* at her mooring in the river at Bundaberg.

Nothing much happened on Monday, but the next day I spent with the 14-year-old daughter of Mrs Hughes, the Scotswoman I told you about last week. Jackie, a cheerful kid, has two very nice horses. Her mother is a Glaswegian and still has a splendid accent, although she has been out here for years; of course I had to meet the family and compare notes on our Scottish heritage. It was a very hot, steamy day when Jackie and I set off, and very soon we and the horses were oozing sweat from every pore. We rode for about four hours, through the sugar-cane fields, and Jackie let me try out both horses, one of which is a racehorse and absolutely gorgeous – Jackie's favourite, naturally! No one else had been allowed to ride him, so I felt suitably honoured. Mrs Hughes met us along the way with ice-cream and lemonade, a most welcome surprise, and that evening I said goodbye to Maryborough – rather sadly – and we drove up to Bundaberg, went on board the boat, motored down to the mouth of the river and dropped anchor for the night. Ian and I have the two forward bunks while Rod and Guy (the final member of the crew) have the two bigger bunks in the main cabin.

We set off at five next morning and were soon in the open sea, rocking all over the place in a fair-sized swell. Even I felt a teeny bit squeamish, though I recovered after a huge breakfast. The first big ship that passed close to us was the *Countess of Jedburgh*, no less, from Ardrossan, so I gave her a special wave. Schools of porpoises appeared at intervals, racing towards us, skimming along just under the bows for a few yards then veering off after smiling up with their intelligent, happy faces. They really are delightful creatures.

The four of us had one and a half hours each steering by compass: a surprisingly tiring job that calls for constant concentration, but Rod did all the chart work and navigation, so we got off pretty lightly in fact. Lunch was cold roast chook and salad and fresh fruit. We did a spot of fishing, catching three beautiful parrot fish, and arrived at Heron Island just as it was getting dark. The island is simply perfect in every respect. I fell in love with it on sight, and would love to live there among its bird-filled trees and coral sand. The reefs around are treacherous to get through, so Rod dropped anchor pretty far out in deep water; consequently we rocked and rolled all night, a feeling one soon gets used to. In fact I can walk the length of the boat with two loaded plates now, which is quite an achievement. Before retiring to our heaving bunks we had a very amusing party, during which we all joined in to make 'tinned curry'. The resulting mess was quite repulsive as curries go, but everyone thoroughly enjoyed it, if only for the amount of effort and thought we'd put into the making of it.

Next day the garbage boat from the little resort on the island came alongside and offered us a lift ashore, as we were not too keen on rowing the dinghy through the vicious little waves and strong currents. The bin-man was a tall, wiry old bloke burnt to the colour of oiled teak, who sat surrounded by glistening cockroaches almost the same colour as himself, apparently unconcerned.

The beach was lovely – silvery white, made up entirely of pounded coral and shells – and the shallow sea within the reef was a brilliant greenish blue. Ian and I set off at once to explore like a couple of 10-year-olds. The huge trees were crammed full of nesting noddy terns, all squawking their heads off. Later we had a super swim in shallow water, among rock pools full of interesting sea-slugs and coloured fish, then returned to the holiday resort, which consists of about twenty huts and a shop-cum-restaurant. There we had an expensive Coke and played the *juke-box* – of all things to find out there on the Barrier Reef!

After lunch we motored out to the reef and fished. I caught four parrot fish in two goes (double hooks), so we had enough for the day; then we shot at floating beer bottles and cans with the .22 rifle. I was congratulated on my aim, even though I didn't actually sink anything. The men wondered where the rifle was kept, and when I said 'in my bunk' they hooted with laughter. I've been ragged ever since for thinking I need to defend my honour by such drastic means.

For tea we had steamed fresh fish with cheese sauce made from peanut oil (yuk!). I thought that was only used for cleaning saddles, but obviously there wasn't any butter. We also ate pumpkin and tatties, followed by fruit; then the two men retired to their bunks, while Ian and I went hunting turtles.

There was a brilliant moon, and soon we found a track on the beach that we quietly followed, only to find three human bottoms sticking up in the air. Three people were closely watching by torchlight one unfortunate female turtle laying hundreds of eggs. She had our sympathy, and we left without disturbing either the people or the turtle.

We left Heron Island this morning at 5 a.m. and have since encountered some pretty rough seas. We've run through several squalls, which made everything rather wet, but now the sun is shining and we are passing another island, just a rocky outcrop with no coral. The porpoises (or are they dolphins? I can never remember the difference) have been super today. One big fellow soared right over the bows and dived smoothly into the sea on the other side.

Christmas Day, 1966

Oh boy, what a trip we had getting into Mackay! After sailing non-stop from Port Clinton all through Saturday and arriving at Mackay about midnight last night, Ian and I woke up clinging to our bunks with all our fingers and toes. The sea was terribly rough and great waves were surging through the forward hatch. I got up to close it and got thoroughly soaked and tangled up with the anchor chain in the process, much to Ian's amusement. We both had a fit of uncontrollable giggles, for some unknown reason, when we found it was impossible to stand upright, as we were beam on to the sea. However, we went back to sleep again without inquiring as to what was happening on deck, waking at dawn the next morning in the shelter of the harbour. We opened our presents with great glee. Many thanks for the humbugs and shortbread. Everyone had a piece of the latter with their morning tea and loved it.

I had a freshwater shower today in the public loos on the waterfront at Mackay, and washed the salt out of my hair at last. Now Ian and I are going uptown for a malted milkshake or two – the cooler the better – if we can find anywhere open on Christmas Day.

Part 2

Meadowbank Station, Mt Garnet

8 January 1967

Two days after Christmas we were still on the boat, and that evening Ian and I took an oar each in the dinghy and had a burn round the harbour, ending up with the giggles and making a racket like a couple of juvenile delinquents. We climbed up onto an old iron barge and thence up onto the far wharf to watch a big American ship being loaded with sugar. Next day we fuelled up and set out for Brampton Island, which is another super place with a holiday resort on one side, surrounded by a clear, calm sea full of little fish. I borrowed Terry's flippers and goggles and spent ages drifting underwater, watching the marine life. Later Rod took *Sue* out further, then cut the engines and we had lunch and shot a couple of empty stubbies, just for the hell of it. On the way home the rope of the dinghy, christened *Sue Too*, frayed; I just caught it before we lost her, skinning my hand in the process and nearly disappearing overboard.

On Thursday Rod woke us up at dawn with the news that the dinghy had broken away in the night and gone out with the tide; obviously she was quite determined to leave us. We got under way, and eventually Ian spotted the wayward dinghy 5 miles out to sea. Returning to harbour for breakfast, we picked up a few friends and took them out to Shute Harbour, Brampton and Lindeman islands. Lindeman was nice, with lots of coconut trees, cockatoos and pet Alsatians, but it does not have a very good beach. We stayed the night moored offshore, and got talking to a lovely family in the next boat, who had come all the way from Sydney with their cat on board and couldn't catch any fish. I went ashore with their daughter to see a film, *The Wonderful World of the Brothers Grimm*, and returned on board to find one of the women visitors dreadfully seasick, Ian fast asleep and twitching like a puppy, and the men still fishing. There was a bit of a slop on, making it rather difficult to remain in bed without hanging on.

Next morning we gave our neighbouring friends some of the fish the men had caught the night before, said goodbye to Lindeman and headed back to Mackay, where we met Jean and Terry, who had come by car. After a Chinese meal we all went to see *The Sound of Music*: definitely the first time I've ever been to the pictures on Hogmanay. I stayed the night with Jean in a motel while Terry took my place on the boat for the return trip. On New Year's Day we collected a party of eleven and set off for Brampton Island, Ian, Terry and I sitting right on the bows hanging on tightly with the waves breaking over our legs. At the island we jumped over and swam ashore in a raging surf, which was rather alarming and possibly foolish of us, but we came to no harm.

That night I had to say goodbye to the Swinbournes, which was sad; they are a really super family. I have an open invitation to go back to them at any time if I'm stuck for somewhere to go, or if I just want a holiday.

On Monday morning I caught the Pioneer Express, a really luxury bus, silent and air-conditioned. At Proserpine everybody piled out for breakfast at a sleepy cafe: stale cornflakes, powdered milk and soggy toast (50 cents). We had a change of driver too. The new one was young and good-looking, and as I was in the front seat we soon got into conversation. I found out that he was an ex-roughrider who had the sense to get out of the rodeo game before he broke every bone in his body, so we had lots to talk about.

At Bowen we ran into heavy rain and the brakes failed briefly, but soon recovered. When we reached Innisfail we were still talking horses and cattle and the driver, whose name was Noel Skuthorpe (related to Lance Skuthorpe, a famous roughrider of days gone by), said I could go on up to Cairns if I had nothing better to do, and he would pick me up in the morning on his return trip and drop me at Innisfail. I made a lightning decision and agreed. After all, I'd have had to spend the night at Innisfail anyway, and Cairns is a beautiful tropical city. Noel was most helpful, finding me a gorgeous motel to stay at and telling me where to go for a good Chinese meal, before he went off on business of his own. I had a good night's sleep in utter comfort and found the big empty Pioneer bus waiting for me outside the motel next morning.

From Innisfail I caught a service car over the Tableland range, through dripping rainforest, along hair-raising roads to Ravenshoe, which was as far as the service car went. There I was picked up in a taxi driven by a Mrs Wellman, the wife of the local mailman, and travelled the final 30 miles to Mt Garnet in style. At the little town I spent the night in a gloomy hotel full of big shiny cockroaches and noisy beer drinkers, and had to be up at 4.30 a.m. to catch the mail truck as it set off on its 500-mile round trip, driven by Mr Wellman.

Many thanks for your letter, received on arrival here. It cheered me up no end, as I have been a bit homesick among so many new people. I suppose it's only natural at the beginning of another job like this; one is so far away from all the people one has grown to know and like over the weeks.

Mr and Mrs Hassall (Bill and Jess) live in the homestead itself. Glen their daughter (27) is married to Ron Rankine, a darkly hand-some six-footer, and they live with their two unusual children, Tammy and David, not far away in a cottage. Then there is Sharon Kruckow, the other jillaroo, who's away on holiday and returns tomorrow. That takes care of the human population. There are two cats called Hennessy and Ptarmigan (the latter looks just like a grouse changing into winter plumage) and a dachshund called Rum. Mr Hassall is justifiably proud of the kitchen garden here; he grows every kind of vegetable, including sweet corn, snake beans, tomatoes, rhubarb, cauliflowers, avocados and pawpaws, and one huge dark green mango tree, which drops basketsful of ripe fruit every day just now, on which we all gorge.

I have done a little housework and gardening so far, also had one short ride to collect a lone cow, and I do the separating of the milk. We have two hours rest at lunch and two smokos of half an hour each, otherwise we work all day from 5.30 a.m. to any time after 6 p.m. Of course I was thoroughly spoilt by the Swinbournes and am finding work rather hard after such a magnificent holiday. Mind you, I have the weekends off, except for separating and getting the milkers up both days.

I have lost almost a stone in weight and an inch round my waist. I weigh 8½ stone now, which is an improvement. Even my un-tannable legs are faintly brown, and my arms and back are a respectable shade (it's not dirt!), but one seems to lose a tan very quickly away from the coast and the salt air.

Thank you for the super calendar, which I have hung in the jillaroos' room. The panoramic views of Scotland look lovely against the white-painted wall. Two walls of the bedroom are made up en-tirely of louvred windows, with a French window on the third wall leading out into the garden, past a sweet-scented, old-fashioned cot-tage rose. A large avocado pear tree stands guard just outside and drops enormous fruit with hair-raising crashes during the night. Our beds are made of stretched greenhide (still with the hair on in some places), with mattresses of horsehair on top.

14 January

The headlines this week are: I have been chased by a buck kangaroo and have had my first Australian 'buster'.

The chase happened last Sunday. After writing to you I went for the cows on foot, saw this large, old man kangaroo grazing alone in a clump of trees, crept up to within twenty feet or so then very foolishly said 'Hi!'. He sat up and stared for several seconds, his eyes changed expression and he began to hop slowly towards me, rapidly gaining speed. I dived behind a two-inch diameter sapling and hurled a bit of basalt at him. He was only about five feet away, and luckily he turned and bounded off. My legs felt like jelly, though I don't suppose he would have done much. Just disembowelled me with his great hoppers perhaps!

The buster occurred on Wednesday, when I had to help Glen get in some more horses. She gave me a real ratbag of an Arab mare called Treasure, who has a small foal. Her front legs needed hobbling before I could get the saddle near her, and even then she hunched her back. I gingerly asked if she was okay and Glen replied indignantly that she was guaranteed to behave and would make an excellent kids' horse. I said no more but climbed on board. All went well until we were outside the yard, then Treasure let fly, just as I thought she would; perhaps I put the evil ideas into her mind. I kept her head up to begin with and survived the first two bucks, but the third sent me over her shoulder and headfirst onto the stony ground. When I sat up I couldn't see for a few moments and had a vivid illusion that I'd done exactly the same thing a minute before, which was odd. However, with a bit of encouragement from Glen – who looked suitably abashed – I got on again. Nothing more happened until we found the horses we were looking for; then Treasure bolted, slowing down only when she ran into the heaving backsides of the others galloping in front of her. Some Treasure!

I have settled in well now and feel much happier. Sharon, the other jillaroo, arrived this week: a cheerful, pretty girl about the same age as me, with curly brown hair and grey eyes. She talks a lot in her sleep. After rather a sticky start I'm finding that Glen and Ron are super people to work for, and they treat me on equal terms. Mr Hassall, the Boss, is dryly amusing with craggy eyebrows and a keen brain. His wife is slightly airy-fairy at times, but kind and considerate and remarkably athletic. All four of them work like Trojans and would never ask an employee to do anything they wouldn't be prepared to do themselves. I don't work quite as hard as I did at Wendouree, can eat as many mangoes and grapes as I like and am allowed to swim in the tank, provided I have a preliminary shower to wash off sweat and dust that would otherwise go into the pool.

On Thursday we rose at 4.30, were saddled and away by 6.30 and collected 400 head of cows and weaners, bringing them home by lunchtime. I rode quite a good black mare called Dewdrop, who looks a bit like Twinkle but is not so absent-minded and nice as that gentle animal.

She's been known to buck badly three times in her nine years, so I'll have to be ready for the fourth session, which is bound to happen when I'm in the saddle. There isn't one completely quiet, reliable horse in all the forty on the place. Oh dear! Well, no doubt I'll learn to ride properly soon.

It is terribly hot today, about 98°F with not a breath of wind. One just oozes sweat, and the flies are awful. Julia Creek today had a temperature of 111°, which would be too hot, I reckon.

21 January

I've done such a lot of wild hectic riding this week that I feel as stiff as a board, but otherwise I'm still in tip-top condition.

The other evening the two jillaroos went to get the cows and found one of them wallowing in the boggy spring. I ventured out into the quagmire, closely followed by Sharon, and wielded a big stick at the bony rump to try and get the old cow (no, not Sharon) moving. However, the stick rebounded and smacked the surface, covering all three of us with warm wet mud, at which Sharon and I fell about laughing helplessly while Cora the cow glared at us obviously offended. She eventually extricated herself with many glutinous plops and ambled off to the gate to join the others, who were all waiting patiently for us ridiculous humans to hurry up and let them through.

On Sunday we all travelled about fifty miles to a station called 'Whitewater' to spend the day with the nice young couple who live there all by themselves. They are terribly short of water; in fact things are desperate. Their one-time beautiful garden is burned completely brown – including the indigenous trees – and dead leaves swirl past and through the house. We sat around all day swigging iced drinks and trying to cheer them up, while the temperature rose steadily to 103°, with a raging hot wind. We returned here at nightfall feeling heart-sorry for those poor people. They haven't had more than four inches of rain in three years.

On Monday I was sent for the horses on Sunny, the night-mare, who is part brumby, part quarter horse, with a decidedly strange mentality. She did one almighty shy on leaving the yards, but was then reasonably well behaved until we began chasing the horses back, whereupon she started galloping and doing tiny pigroots (bucks), getting faster and faster until we overtook the mob we were meant to be driving. I couldn't think what had happened to make her go so silly, until I glanced down at the flying ground and saw the end of the girth strap trailing. She was treading on it, and the more she trod on it the more frightened she became.

Tuesday was a more successful day. Sunny and I ran in the others again, yarding them without mishap, then later I went out mustering on Dewdrop, who was impossible for the first hour. She fought for her head all the time until she ran straight into a sapling, which she couldn't see because her nose was pointing up in the air. We were left to put a cow through a wire gate while everyone else went to view a new foal a few yards away. After opening the gate I remounted, but before I could find the offside stirrup she was away, heading flat out for the group around the foal. Later the same day we had trouble with a frantic Brahman bull who charged everyone in turn, so once again I was galloping.

Next day we were up for 5.30 breakfast and spent the whole morning branding about three hundred calves. I had the job of keeping the brands hot with a flame thrower – as opposed to the fuel-drum fire at Wendouree – and handing each iron up to the bloke who was doing the branding: 'ZOS' and the year number. I also got the horses in just before morning smoko time when the mail arrived – a big event here, as it only comes once a week – then after lunch we weighed the weaners. In the evening Don and Dell Lavers turned up, friends of the Hassalls, and quite a party developed on the verandah, helped by a bottle of Kaiser Pearl sparkling wine.

On Thursday Sharon and I were up at 4.50 and had breakfast at the cottage with Glen while Ron got the horses. I rode Dewdrop again, this time with a curb, and she was quite good all day, helping to bring about three hundred cows home by lunchtime to be pregnancy-tested in the afternoon. This entailed a lot of chasing around from yard to yard, a scary kind of job as those big fat Brahman matrons can be quite fierce when they feel like it. I read out each ear tag, kept the crush full of angry cows and generally raced everywhere getting covered in black dust and sweat. Storms raged around but no rain fell until six that evening, when it poured down and gave us 80 points of welcome water.

Yesterday we finished testing. The remaining cows were very toey, and I was charged three times; I'm becoming quite adept at leaping for the top rail from a standing start. The non-pregnants were taken out to their paddock first, then we returned for the ones in calf, who went elsewhere, and for this work I rode a brown mare called Dawn. She's very nice, though a little hard to hold when she wants to be off.

After a late lunch Glen, Sharon and I set out to find some fresh horses in Billygoat Paddock, galloping here and there. At one point Dawn jumped over the road, which was unexpected but gave one a nice floating feeling for a few seconds while she was airborne. She was going too fast for me to do much about it anyway. It was late in the evening when we returned; then Sharon and I had to take our two mares and lead another one to a separate paddock away from their foals. I led Dusk from Dawn (at dusk), and everything was fine until we reached the gate, whereupon Dusk pulled away in panic and raced the 2 miles home trailing the rope, with Sharon in pursuit. I nearly fell off Dawn, as I was holding the rope in both hands. Dawn was by then terribly excited and wanted to gallop home too, but I somehow made her trot fairly sedately, arriving just in time to see Dusk crash into the barb suspension fence and fall down. She cut her face quite badly, and it was a very subdued mare who finally allowed herself to be taken back to the yards.

Today there was no riding because we had to do a large amount of washing. Tomorrow we are going to muster one of the southern paddocks, camping out for at least three nights and taking three horses each. Help! It should be good fun actually – if I survive.

Mr Hassall is most interested in my background and is forever asking questions about Scotland and Ceylon. Last night during supper he told me he's going to write to you, Dad, as he reckons you'd like to know what your daughter is up to.

We have superb food here: masses of vegetables from the garden, lovely meat, tatties in their jackets, loads of homemade butter, cream and milk, honey and brown bread, not to mention all the fresh fruit dropping off the trees, including cumquats – small bitter oranges that make a refreshing drink.

I had news of Don Hafemeister the other day. He's broken another two ribs down at some rodeo in New South Wales but has been winning a lot of events, and so has my mate Dougie.

29 January

In answer to the queries in your last letter: my housework consists of washing-up, filling the fridge with kerosene, a bit of rough sweeping, making biscuits or a fruit cake for smoko time, and cooking the evening meal on alternate nights. Otherwise I work in the garden watering, mowing, picking up mangoes and sweeping leaves. The cattle are beef animals, not dairy cows, and are a cross of shorthorn or Devon red with the extraordinary-looking Brahman, which after numerous generations eventually produces a Droughtmaster, a breed well suited to the hotter parts of Queensland.

The names of the various paddocks are: Billygoat, Junction, Bowerbird, Moneymusk, Glendhu, Top Brumby, Bottom Brumby, Spring, Wabble Creek, Beasleys, Alphabets, Bull, Scrub, Big and Small Weaner Paddocks, The Square, Jacks and the Horse Paddock around the house. None of the fences runs straight. They all have about six different angles in each one, which makes getting around rather difficult; I usually carry a map.

Yes, I am sharing a room with Sharon. She is 19 and engaged to a bloke called Angus. By the way, the boss has written to you, but never having used an aerogramme before, has put stamps on it. In all fairness it should go by Blue Streak.

Now, the camp. I must say it was thoroughly enjoyable, hard as the work was; I seem to have found my vocation at last. On the Sunday Ron and Sharon took the horses down to Wabble Creek Paddock, which is some thirty square miles in area, while Glen and I took the truck and gear down and put up the tucker tent under which all the boxes of food, etc. were stored – off the ground because of the white ants. We also erected the two smaller sleeping tents in among the tall, thin gum trees, prepared lunch and ate it, then the four of us did a little gentle mustering in the afternoon. We found 229 head and put them into the small holding paddock beside the camp. I rode yet another black mare called Shaneen, who is nearly 16 years old and the fastest walker on the place. She's a dear old thing with a pleasant face, but her steering's a bit faulty; she tends to crash into rather big trees while cantering unless given clear instructions well ahead.

We returned for supper at sundown: steak, onions, choko, snake beans and tatties baked in the fire, cooked by the light of a carbide lamp spluttering on a tree stump near by. Then we sat around the fire and watched the full moon come up, while Glen played her accordion. She's self-taught and plays quite well. That night we spread our swags on the ground, having dug a depression for our hips, and

I had just settled down on my three blankets with only a sheet covering me when Sharon gave a strangled squeal. She'd just discovered a 6-inch glossy brown centipede marching across her pillow straight towards me. As you can imagine there was an uproar in our tent until the wretched thing went to earth beneath my socks, whereupon I guided him out under the canvas flap and hoped fervently he would continue in the direction he was going. Subsequently I slept remarkably well on the hard bed, though my feet were nearly eaten away by the mozzies.

Next day I was introduced to Bits, a large, light bay mare whom I was to ride. She has an unfortunate habit of sitting down promptly if one pulls her up on a tight rein too quickly – which of course I did. She sat down, I slipped off and she rolled over onto my long-suffering left leg, the knee of which hasn't fully recovered yet. After being given a leg up again (as it was too painful to mount and Bits is over 16 hands high), we got on very well, accomplishing a lot of good work throughout the day.

Next day Ron's horse became extremely lame, so he had to walk back to camp while I was left to take a mob of young steers back to the holding paddock alone. It was quite alarming but we got there intact.

On Wednesday I rode big Bits again, to take all the cattle home together with the spare horses. It was a heck of a job keeping the mob moving; our voices got drier and crankier as the day wore on. It took six and a half hours to get them home, after which I drank four mugs of water, three of separated milk, one of cold coffee, two of tea and ate six mangoes. The cattle were then sprayed for ticks, drafted and taken back to their new paddocks, and we eventually finished by 7 p.m – a splendid day's work.

I rode Shaneen the next day, who behaved perfectly, walking all the way home like a good soldier until we reached the shed, where she took off round the corner and skidded to a stop at the unsaddling rail, whinnying thankfully. I couldn't help laughing at her obvious relief – I'd swear she was smiling. In the yard work that day Glen had to cut the tip of one poor thin bull's tongue free from his bottom jaw, as he was tongue-tied, literally. It was a horrid job, but the bull should be able to feed properly now and put on weight.

Friday was another day spent in the saddle, mustering bulls to bring home for spraying and sorting out, then returning them to different paddocks. They are quite terrifying in the yards; I'm afraid I spent most of my time on the topmost rails while they prowled and rumbled down below like a pit full of tawny lions. Yesterday there was a riding job to do too; we certainly had a good week's horse-work.

There is a cyclone just off Bundaberg at present, rapidly approaching Maryborough, and the Mary River is nearing flood level already – they had four inches of rain yesterday, according to the radio news. I hope *Sue* is safely moored. I had a letter from Jean, and one from Ian back in Sydney, where he has just joined the Navy. They said Pop and Terry, on their way back from Mackay, had been blown onto a reef at Heron Island but luckily the boat suffered no damage.

Today is a really nice Sunday off. We were allowed to sleep in until 6.30 and didn't have to milk or anything. It is great to have a day completely free: it makes such as difference to one's outlook. Sharon and I climbed Meadowbank Hill behind the house (it's an old volcano) this morning, coming across a goanna, a snake and lots of rock wallabies hopping about. There was a lovely view from the top.

Smoko time has arrived, so I shall stop rambling and be sociable over a pannikin of tea.

Sunday, 5 February

Towards cow-time on Sunday big clouds gathered and rain poured down, so I collected the milkers in my bathing togs once again and swept out the six horse stalls while waiting for the calves to finish their evening drink. Mrs Hassall has told me they will pay me $16 per week from the start, instead of the $14 as agreed for the first two months. They've employed girls since 1930, and apparently yours truly is the most promising jillaroo they've ever had. Now that is nice to know – let's just hope I live up to their expectations.

Next day I fetched the fifteen horses in at eleven, then after an early lunch we set off for Beasleys Paddock, 14 miles away, to camp for another three nights and muster during the days. Glen and I took the horses down; I rode Sunny, the only one who is more or less safe behind horses. She's a lousy walker, though, and needs constant coaxing to move forward at any speed. We reached the camp two hours later to find that Ron and Sharon had erected all the tents and were agreeable to preparing supper if we wanted to scout around for cattle, seeing that we were still mounted. Glen and I spent a pleasant hour or so in the nearby bush and came home with seventy head.

That night after sundown we sat by the fire and sang songs to Glen's accompaniment on the accordion. The 'Skye Boat Song' sounded a bit incongruous in that hot, sub-tropical environment; nevertheless it was rendered with great gusto, along with others like

'Yellow Bird', 'You are My Sunshine' and of course 'Home on the Range'. I slept very badly that night. The mozzies were awful and the heat was even worse, so that breakfast came as a relief: eggs cooked in leftover gravy together with fried onions and old vegetables. I rode Bits that day, but she must have got out of bed on the wrong side for she was in a rotten mood and when we took off after a steer she just about ran the wretched beast over, her ears clamped meanly back along her skull. She's such a *big* horse to control.

After lunch I went out with Glen. We trotted, trying to avoid the showers of rain, but got soaked in the end. Bits started jogging as I was trying to sit well down in the saddle and keep a dry bottom, so even that got wet, and both boots overflowed. Thunder crashed, lightning flashed and when Glen asked me to hold the slack fence down so that she could jump her horse over to collect a stray beast on the other side, I got an electric shock from the wires; lightning must have struck it somewhere further up the line. It certainly gave me a fright, and Glen too, for that's the way a number of horses and cattle are killed by lightning.

Back at the holding paddock we had awful trouble with three bullocks who would not go through the gate, and Bits was *vile*, going backwards, sideways and up and down, sometimes forward with her head down between her legs, all on very wet, greasy mud. I gave up trying eventually and let the others push the cattle through. It rained again in the night, but I slept well through sheer tiredness and next day went out on Shaneen, who was sweet all day and so eager to oblige. Ron went one way and I another, having agreed to meet at the mill. Shaneen and I found only one cow all morning but duly took her to join the 250 waiting at the mill with Ron, Glen and Sharon, who laughed at the enormous herd I had condescended to bring along. That particular cow, a brindle Brahman type, was called Mrs Potter, as she had a certain likeness to the bank manager's wife in town. I wonder what the original Mrs Potter would have to say about that!

Sharon and I took the mob back to camp in time for lunch at 2.30 p.m. We didn't go out again that afternoon; instead we had a welcome siesta on our hot swags for an hour, then I made a rice pudding. I defy anyone to make a successful rice pud in a billy over an open fire, but it turned out fairly well, if a bit stodgy.

Glen and Ron returned later with more cattle. I'm becoming very fond of those two; they are really fine people. Glen is thickset and as strong as a man, but vivacious, kindly and amusing, with a fierce pride in Meadowbank. Ron is fearfully capable at everything he turns his hand to and has a wicked sense of humour. We had a hilarious night round the fire once again; the jokes were flying thick and fast and we laughed until we cried at the stupidest things.

Next day saw us on the move before 6.30 a.m., having packed up and loaded all the gear onto the truck. Glen and Sharon went off to muster, leaving Ron and I to deal with a herd of 640 cattle, who became impossible to control as soon as they realised there were only two people driving them. We had to let the horses go on ahead by themselves, as it was too much trouble to try and keep them with the milling cattle, but we caught up with them waiting at the top gate to be let through. It was lunchtime before we got them all home safely; then I had to run in fresh horses on Sunny, as Sharon and I were to go out and find four steers in Billygoat Paddock.

We found them all right, but they were scrubbers of the first water; they split in four directions at full gallop, so we ended up taking them home one by one, flat out. I tried to race the first one back to open the home paddock gate, but he beat me to it, sailed through the top wire of the fence and disappeared towards the yards.

That evening the government vet, Peter Darvall, arrived to inoculate some heifers the next day; then on Friday Ron, Sharon and I took some weaners out to Wabble Creek, which took five hours. We came back, put ten head in another paddock, did a bit of drafting in the muddy yards and once more took some heifers out, so poor old Bits was in constant use from dawn until dusk – and so was I.

It rained all day yesterday: monsoon-like weather, with grey ragged clouds drooping over everything. I rode Shaneen to take the last mob out of the yards to their new paddock, and she slipped and skidded all over the place on the black, greasy soil. With all this riding I seem to be getting a little more confident in partnering these splendid stockhorses, though they still take me by surprise every so often, and I doubt very much whether I've had my last buster.

12 February

Ron was in town last night and brought your letter back with him, for which I thank you. No, I won't chase any more kangaroos, at least not on foot; I should have known better, but was just curious as to what he would do.

This week has been soggy and grey with the onset of the wet season. It's made mustering a bit cooler, but all the same I prefer the sun. There's been no rain worth mentioning, just miserable drizzle, which does nothing except make one damp, the reins slippery and the horses ill-tempered.

On Sunday afternoon I watched Glen and Ron making sausages, which turned out to be an amusing occupation, especially when they made me pull the end of the dried, salted pig intestine over the garden hose and turn on the tap to stretch the skin prior to filling it with mince. I always thought sausage skins were made of some rubbishy kind of plastic. One beast killed here lasts from six to eight weeks, and they use every scrap, even down to the feet for brawn and the brains and central nerve for the children.

On Monday I went fencing with Mr Hassall and the children, Dave and Tammy. I took turns at driving the Landrover, an alarming vehicle quite unlike the Wendouree one, with the most awful steering – something to do with a tie rod, apparently. Then next day we mustered another paddock. I rode Bits again and she behaved perfectly all day, making me wonder whether it was my fault she was in such a bad mood the other day. It drizzled most of the time, but we got the mob of about 200 in by lunch, then had all the drafting and spraying, etc. to do. There were six little weaners with terrible poisoned dingo bites. I had to hold down the head of each one while Mr Hassall slit open the suppurating lump with his penknife to allow the filthy yellow pus to drain away. The stench was revolting; I'm not surprised graziers loathe dingoes as much as they do.

In the late afternoon Sharon and I took the mob back to their paddock, getting home after dark, some time after 7.30. Some days we work fourteen hours or more, but they are a very fair crowd here, and every hour of overtime will be repaid with holidays, etc. Yesterday I received my first cheque for January: $67.20, which isn't too bad really, especially as it is mostly pocket-money and I'm living with nice people, with lots of time off and wonderful food. I only wish dear old Twinkle were here too.

Sharon and I had braised steak at the cottage on Wednesday at 5 a.m. (Usually breakfast is provided in the big house, but when there's an early start we have it with Glen, leaving the milking for the Boss to do.) I rode Shaneen, and we mustered one of the more northern paddocks that has the airstrip in it. Unfortunately the mob of weaners decided it was a racecourse and took off like a Yankee stampede, galloping the full length of the strip pursued by the horses, who of course enjoyed it no end.

At lunchtime Sharon's fiancé Angus arrived, having just received the sack from his previous job, and is now hanging around here hoping for work. He seems a pleasant enough sort of bloke, though not exactly 'glamorous' in any way. I rather get the feeling that Sharon isn't all that enamoured with him either and that his arrival here has embarrassed her somewhat. Ron and Glen are dying to see how the situation develops.

One of the milkers calved mid-week: the pretty little Jersey called Gumnut, who had a heifer that is being called Gumdrop. I won 50 cents, as I bet she would be the first of the three milkers to calve.

We jillaroos were up again at 5 a.m. on Thursday, packed our lunches and after an early breakfast we all set off with the rising sun to muster three relatively small paddocks. There was a lot of preg-testing to do that day, and Peter Darvall, the vet, came to do a post-mortem on a poor rickety bull, who charged everyone with a dazed expression on his thin face, until Ron shot him and dragged the carcase to the gallows with the 'rover. I wonder what was wrong with him.

On Saturday a buyer and two mates came along to buy some bull-ocks. They stayed for lunch, much to Mrs Hassall's horror, as she had very little of anything ready. Between us we made various salads, for which she was very grateful. There was too much food in the end.

Bits is going to a stallion some time in the near future, when Ron has finished the crate for the back of the truck. The stallion is on a place called Spring Creek, about sixty miles away, owned by Mrs Hassall's sister and family. I have been invited to go along, mainly I suspect because there is a son of 27 there. All this matchmaking that goes on – really!

Are you snowed up or is it nearly spring? Here it is cold and blustery and I'm wearing a jersey – so much for the tropics. In fact there is a suspected cyclone developing in the Gulf, so we might get some proper rain from that.

18 February

Becoming bored on my last day off, I went along to help Ron and Angus finish painting the crate for the truck and cleverly managed to fall right through the bottom, receiving a long graze on one shin and a cut on the other. I didn't drop the paint tin, though. The others fell about laughing, including Glen, who came running out of the cottage to save the *paint*. There is something about the Aussie sense of humour that appeals to me, though I can't think why.

Next day we had early breakfast and went out to muster one pad-dock. I rode Bits, helping to put all the cattle to one of the bores, where I was given the job of holding the mob together while the others cut out sixty high-grade Brahmans and eight bulls. This camp-drafting is interesting to watch, as each beast that is cut out of the herd can run, head up and tail vertical, as fast as the average grassfed horse. Bits was fairly good, but she is such an unfriendly horse. She just gazes into space and bites you if you attempt to stroke her. She also heaves great depressing sighs straight from her heathenish soul at intervals, just to make sure you're feeling guilty for having saddled her that morning.

The cows and bulls that had been cut out were put into another paddock alone and we came home to learn that we were going to Cairns the next day. Sharon and Angus broke off their engagement that night, which was a relief for everybody, but he is staying on here to help Ron build a new hay shed.

Anyway, we left at seven on Tuesday morning, Angus and the kids perched on the back of the truck, Glen, Ron and I in the front. It was a pleasant journey down over the lovely Tableland, stopping for lunch in Atherton, where I had chook and proper spaghetti with cheese on top. At Mareeba we shopped a bit then went on over the glorious approach to Cairns, a breathtaking range of rainforest-clad hills and terrifying hairpin bends.

The Rankines very kindly asked me to share their room at the motel, as Ron was going off fishing anyway, so I accepted gratefully, paying for myself. I rang Jean Swinbourne and surprised them all – it was good to hear their happy voices again. Glen and I had a lovely dinner of devilled chook, Tahitian fillet steak in wine and rice, then coffee as we had no room for a pud. I slept quite well, although I don't take kindly to air-conditioning, as you know. Mind you, Cairns is pretty humid – similar to Colombo in fact, as far as the weather and riotous flowers go – so some sort of air-cooler is a good idea at night.

In the morning Glen and I and the kids boarded the tourist boat for Green Island and got there by ten. I went out in a glass-bottomed boat but was a little disappointed in the colour of the coral, though there were lots of pretty fish. Later we visited the Underwater Observatory at the end of the jetty: a great piece of engineering that enables you to look through different portholes at the marine life on the seabed. We could have stayed there a long time, but Tammy and David were hungry. That night I took Glen out to the Kowloon where we had ourselves a terrific Chinese meal. She agrees that everyone should swap dishes so that each has a taste of what the other is having.

On Thursday we parked the kids in a day nursery, with some apprehension. (In fact they loved it once they got used to the idea of hordes of noisy infants.) We spent the day shopping and browsing happily until two, when we went to watch the *Possum* come in with Ron and his mates from the fishing expedition. After a wash and brush-up we all went in a big party to the Kowloon again, thereafter to see *The Sound of Music*, which I enjoyed just as much the second time.

We left Cairns the next day; it's a lovely city, hot and full of interesting smells. Ron went in the truck to load trusses and timber at Mareeba and pick up Angus again, while Glen and I took the kids on the funny little diesel train that goes up the Tableland via Kuranda. Twice the engine almost came to a stop, once tooting its whistle at five turkeys on the line and again when a swamp pheasant perched on the track. We thought it most considerate for a mere machine to slow down for the local bird life. There were fabulous waterfalls and gorges far below the swaying train as it clattered over rickety bridges on the narrow-gauge line. It rained continuously all day, and at Kuranda we raced out and bought soggy stale biscuits and some mouldy chocolate, eventually reaching Mareeba to meet up with Ron.

After eating some very good fishcakes in a cafe, we all drove up to Ravenshoe, where we learned that over an inch of rain had fallen at Meadowbank and we weren't to attempt the trip back. We spent the night with some of Ron's relations; it was rather an invasion, but they didn't seem unduly perturbed – neither did their four nice cats and one ancient, smelly, black retriever. We slept until 8 a.m. the next day and set off in leisurely fashion, travelling very slowly, with all the makings of the hay shed on board, and arrived at 3.30 p.m. It's good to be back, but I'm very fond of Cairns. It's so tropical and colourful, surrounded by high, tree-covered hills and ragged clouds that throw navy blue shadows.

27 *February*

Last Sunday we took the three mares – Bits, Dusk and Flax – over to Spring Creek Station, having lunch at Lava Plains on the way. The quarter horse stallion named Noble Don was standing in the yards as we arrived, a splendid, dark liver-chestnut beast, all rounded and muscled and arched. We sat down to tea with the Collins' family, but had hardly broken the ice with them before it began to rain hard and we had to rush off to try and get back while the road was dry. I went in the truck to open gates while Glen and Sharon came behind in the 'rover, but the latter broke down with wet points, so Ron had to turn the truck round on the narrrow black-soil road and go back to help. Eventually we got home at 8 p.m. after sliding around all over the place, missing trees by fractions and at one point even considering abandoning the truck – until we found that just a few yards further on the road was bone-dry.

Monday started my week of milking, and do you know, by the end of the week I was getting a *full* bucket. I also cut my hair, much to everyone's surprise. They consider it looks too 'bangtailed' now and are threatening to feed me on spinach to make it grow again.

On Wednesday Sharon got in the horses so that she, Glen and I could take the dried-off milkers and a batch of poddy calves to a distant paddock. I rode Shaneen, who was surprisingly fresh and kept trying to turn for home, even when we became involved with chasing two bulls away from a fence where they were making eyes at some heifers on the other side. That night we went to supper with Glen and Ron at the cottage and ended up doing crab-walk on the floor, much to the astonishment of the dog, Rum. He raised his ginger eyebrows very disapprovingly at all the gymnastics and took himself off to bed in disgust.

On Thursday I mowed from 8.30 a.m. to 5 p.m., with a couple of hours off for lunch, and made a splendid job of the lawns, though I say it myself.

The Hassalls have that delightful book *The Specialist*, which they all love; in fact the 'dunny' at the big house follows exactly one of the designs (for a one-holer), even down to the carved airhole in the door. They keep stacks of the *Cairns Post* and *Queensland Country Life* for folk to read if they are to be in there for any length of time.

Ron shot a dingo that has been coming in every night to drink from the trough and chew at the old hide hanging on the fence at the yards. The moon was full that night, lighting up the ground well enough to let Ron shoot from a fair distance away. The milkers kicked up a terrific fuss as the dog approached their locked-up calves.

I have spent a bit of time pruning a couple of orange trees and tidying up the chook yard where the cumquats grow. The mangoes are finished, sadly, but the guavas are starting, though unfortunately they are the large white variety that aren't much good unless stewed – the seeds are a bit powerful. Also, I have started collecting caterpillars again. (Dear me, I must be approaching my second childhood.) I found a beauty yesterday, though, which needs attention now. Dave and Tammy are entranced by the way it continues to eat while kept in a cardboard box and spend ages hunting for the right kind of leaves for it. They really are nice kids, only 4 and 2 years old, but one can talk sensibly to them and show them things in which they take a great interest. They've taught me quite a few things too.

I am writing this sitting under the mango tree, and the day is just perfect, with not even many flies about. Hennessy is keeping me company. I tried to make Ptarmigan sit for a photo under the bougainvillea, but he objected strongly when Ron started up the truck near by, scratched my arm and stalked off. Now he isn't speaking to anyone – a cranky cat, or maybe even a peeved pussy.

4 March

It has been a short week, last Monday being my day off, and I'm having today off this week as Sharon is having tomorrow off. A buyer is coming on Monday with semi-trailers to collect some bullocks, which we will have to muster then, that is if it doesn't rain in between, which it did last night and gave us three inches; so you see it all depends. (Sounds a bit complicated but it's not really.)

On Tuesday last I was loaded up with a 2-gallon spraying machine and, feeling like a pack-horse, set off to spray the guava tree for fruit fly, also the cumquat trees and the Chinese orange tree. Then Mr Hassall wanted me to type a long letter for him. The next day was mail day, a chaotic hour with parcels and boxes everywhere, kids running off with bills, Mr Wellman, the mailman, trying to eat his breakfast and the dog barking, in the middle of which I was making Anzac biscuits, chocolate cookies and junket (at 7.30 a.m.). The rest of the day was spent weeding and generally messing about. Thursday I started cooking early again – 8 a.m. – and for morning smoko made two lots of biscuits that disappeared within minutes. Then I killed young banana suckers and picked some Japanese passionfruit, some cuttings of which the Boss says I ought to bring home with me as they survive frost quite well. They taste like ordinary passionfruit mixed with cream. Yesterday we had curry for breakfast and I made *more* biscuits and a huge fruit cake. After that I was put into the new crate, given a knife and told to scrape the dried dung off the wood on the floor and then paint it all with sump oil. Surprisingly enough it was a most enjoyable job. I got it finished by afternoon smoko, when I made fishcakes from your recipe, under the mango tree. We had them for breakfast this morning, much appreciated by all.

Last night I went down to see Ron and Glen after supper for half an hour. It started to rain and I had to come back to the house wearing Ron's enormous welly boots and carrying a golf umbrella. I nearly got swept away. Sharon and I spent a disturbed night with all the louvres shut as the rain seemed to be coming in from every direction. Everyone is jubilant today, especially the Boss, who raced out in the Landrover to inspect all the dams. The way the 'rover came out of the shed (broadside) it's doubtful if he will be back for some time, but he did take a shovel. The ground is just like an enormous bog with trees growing in it.

The Swinbournes had their back balcony demolished in the cyclone Dinah a few weeks back, also a couple of young trees, but nothing too serious, although Jean said it was pretty terrifying. Poor Maryborough has had its share of the elements this year, what with that awful hailstorm and now Dinah, and they have had more rain than for many years.

The next time there is any mustering to do I shall be riding a mare called Damona. Actually she's supposed to be a steady, reliable old thing whom Mr Hassall rides when his own mare Echo is tired, but he hasn't ridden Damona since the time she pulled away from him and galloped backwards across the yard until she sat down. That, according to Glen, is the worst thing she has ever done, so perhaps she will suit me; but then of course I seem to have an odd effect on horses, so one never knows. Glen isn't saying anything yet. She's a most unusual person, who says exactly what she thinks – truthful to the point of bluntness. She's a splendid soul and I'm very fond of her. Ron is nice too; he's very amusing and looks like a tanned version of Sean Connery as he was in *Dr No*.

12 March

Many thanks for your letter received ages ago. I am *not* a wiry old horse! The ABC told us today that Britain is having snow and gales, so I hope it isn't too bad with you. The wet season has arrived here with a vengeance – the best they've had for years. (Rumour has it that *I* have brought rain to all these drought-stricken areas of Queensland!) The Hassalls have gone to Cairns for a week and are due back on Thursday, but last night Cairns had 16 inches of rain and all the shops are flooded with 3 feet of water, so we have visions of them sitting on top of the fridge in their motel room watching the bedding float by. Obviously they won't be back for a while; in fact with the rain falling here too I doubt if we'll see them for months. Sharon went home with them, so the cats and I are occupying the house in solitary splendour. Meadowbank has had 11 inches of rain this month and a couple of dams are overflowing, so everyone is quite pleased to be housebound for a while.

The Boss says I'm to tell you how the Droughtmaster came about, so here goes. In 1906 some Zebus were imported into the USA from India to form the basis of the Brahman breed, which is a mixture of Zebu and British shorthorn or Devon red, so a Droughtmaster is a

cross between Brahman and shorthorn mostly. The Society book*
says, 'Droughtmasters are a mixture of Brahman and red beef cattle
and may have from approx. $\frac{3}{8}$ to $\frac{5}{8}$ Brahman blood. The higher per-
centage Brahman seem to be better for the tropics. Culling for
conformation is of course very necessary, but so also is culling for
temperament, (ha ha!) tick resistance and quick growth'.

Well, last Sunday three scientists from the Department of Stock
arrived in a little, chartered one-engine plane that we all met
at the strip. It absolutely bombed in, and I was sure it wouldn't be
able to stop before the end of the tiny runway in the middle of the
bush, but of course it did. While the men went off to look at cattle,
Sharon and I had a lovely swim and she started to teach me to dive.
Ron reckoned it was all wrong, however, so at six o'clock last night –
in pouring rain – he dragged me out and taught me another way.
I still can't dive.

There was a party on the verandah for the visitors, which was most
enjoyable ; I felt more like a member of the family than one of the
hired hands. One of the guests, called Dave, offered me a job in the
Kimberley mountains of Western Australia, mostly riding, but it
seems a long way to go so I declined.

On Monday I clambered aboard Sunny to collect the seventeen
other horses, who fairly galloped into the yards with Sunny bringing
up the rear, shaking her head like a mad thing. Later I rode Damona
for the first time, and went to bring home a lone bull from one of
the paddocks – a nearly impossible task, as he headed off in all directions and almost ran us down trying to get back to the cows through
the gate. My hands kept becoming entangled in Damona's greasy
black mane, so as soon as we got home I found a pair of scissors
and cut it all off. She now resembles a rather moth-eaten clothes-
brush.

The visitors left that evening – just in time, as 2 inches of rain fell
half an hour after the little Cessna took off. During that rain two
most peculiar people turned up in a car and got themselves bogged
at the home paddock gate: a woman of about 25 and a boy of about
19 with long black sideburns. They had a huge white *turkey* sitting
on the back seat. The Boss reluctantly pulled their car out with the
'rover, but they broke down 9 miles up the road and returned on
foot, as they had no tools at all. Everyone here was pretty mad with
them; it was a topic of conversation for quite some time.

I made bread the other day for the first time ever. It took all day
but turned out quite well; at least it all disappeared, and I don't *think*
it was fed to the chooks. We're all having to sit around doing nothing, and we shall get terribly fat. For lunch today I made a huge pan

*The Droughtmaster: Australia's Foremost Tropical Breed Bred by Australians for
Australian Conditions*, The Droughtmaster Stud Breeders Society, Brisbane, 1965.

of fried rice, added asparagus, tomato, herbs and hard-boiled eggs, and put strips of freshly grilled steak on top. *Very* nice, though I says it. Sharon and I went to supper with Ron and Glen, then the four of us played Drawing Consequences. We had some really hilarious results almost worth framing. We laughed a lot and didn't get to bed until about 10.30, two hours later than usual. Mr Hassall and I had an argument about the Loch Ness monster that was threatening to become really heated when Ron chipped in and called Nessie 'the Scotch Croc', which made us all dissolve with mirth again. They are a good crowd. Angus is still here, but not for much longer as the hay shed is almost complete.

The mosquitoes are here in full force now, breeding in every ditch and puddle, and there is a continuous whine about the place. And yesterday – oh dear! – I had the most ghastly experience. I still shudder at the thought. I made a cup of cocoa and there was a lump of what looked like unmelted chocolate on the top, which I tried to stir down but without success. I drank half the contents and was about to chew the lump when I realised it was a *small cockroach*, still alive, though a bit bruised and soggy! Ugh!

Monday It's still raining. Three inches fell today, and the dam just above Garnet is threatening to engulf the town with its 6 million gallons of water. Some stations have started evacuating, and some are bogged and in trouble. We have been listening in on the two-way radio; it's all most interesting and exciting, but I do hope nothing awful happens to anyone.

Thursday Communications are almost nil, only the transceiver now. There is 14 feet of water on the bridge at Mt Garnet, the phones are all out of order and of course the mailman couldn't get through yesterday.

18 March

I'm still writing weekly letters, although the last one hasn't gone yet. We've had a hectic time here, though the floods, which have done a dreadful lot of damage elsewhere, somehow seem to have missed us.

Last week we had 2 inches of rain on Saturday, $5\frac{1}{2}$ inches on Sunday and 4 inches on Monday, but now it seems to be clearing up. On Monday we heard over the transceiver about the awful flooding –

14 feet over the Garnet bridge, and the tin dredge dam about to burst, which it did that night, washing away the whole of Glen Eagle Station. All we heard of the Blakeneys was that they were evacuating, then nothing more until Wednesday, when I was left in charge of the hateful transceiver. Someone rang me on the phone and said the Blakeneys from Glen Eagle had just driven up to their place with a swag each and the Landrover; everything else had gone. Mrs Blakeney was a thyroid case and her pills had been swept away, so I had all this to tell VKA the Flying Doctor on the radio. Never having used it before I was in a state of near panic. I had to learn as I went along, saying 'Roger' and 'Over' and pressing the right switches, while all the time everyone was hurling questions at me and wanting me to relay telegrams back through the phone. The static was awful and I could hardly hear a thing at times. I took as much as I could in shorthand, but my hand was shaking so much with sheer terror that it was nothing like the original Pitman's I learnt in Elgin. I had to order a helicopter to drop supplies, then rush back to the phone to ask what their strip was like, what was the visibility and so on. The kids had been left in my charge for the day, and in the middle of an important cable, piercing screams came from the nether regions of the house. I had to hastily interrupt the caller, and rushed off to find Davey hanging by this fingertips from a manhole in the ceiling, as Tammy had taken the stool away. It was she who was screaming for me to rescue her brother; it didn't occur to her to put the stool back herself! So it went on all morning, and by midday I was almost in a state of collapse.

What had happened was that the day before, Ron, Glen, Angus, the kids and I had gone in the tractor to see the new dam, which has never had much water in it. When we got there Ron drove straight into a huge hole, where the tractor sank rapidly, but everyone was so pleased to see the dam overflowing that we abandoned the machine with hardly a thought and raced up onto the dam wall. Needless to say we walked the 6 miles home. The next day, Wednesday, the three grown-ups went on horses with wallaby jacks, etc. to dig out the tractor, leaving me to listen to the transceiver, with instructions 'not to talk unless you have to' . . .

Hey, I've just been to the phone and got your telegram. Thank you for bothering to reply. No, we haven't had any damage bar a few broken fences, which Angus is out mending today. Our only complaints are about the shortage of mail and the fact that I have to make bread and dampers non-stop. But poor Glen Eagle. Ron is over there at present helping to salvage things and build a shed for the Blakeneys to live in. Their car is a write-off, the cattle truck has been badly damaged, foals are drowned and the majority of their cattle swept away or hung up in trees.

Anyway, after that frantic morning, which only eased up when Glen returned to my rescue, the kids said they wanted a ride. I rode Shaneen and led an ancient little pony with Tammy on board, while Glen led Davey. All went well until we turned for home. Shaneen usually needs two hands to hold her back, as she tends to move rather fast in the homeward direction. Tammy almost slipped off her pony, I lunged over and grabbed her shirt before she disappeared and the four of us galloped into the yards, which were luckily near, in a horrible tangle of leading reins and flapping leather. Altogether a nerve-racking day.

The Hassalls are still in Cairns, intending to fly back soon, but the strip has running water alongside it and won't be fit to land on. Today is the first time the sun has been out for about ten days; it is really good to see it.

Sunday I am working today as the two daughters and son-in-law from Glen Eagle have arrived and are staying the night. They'll be collected by a 4-wheel drive tomorrow when Ron comes back. Poor souls, they are so dejected. Imagine having nothing but a shed to go back to in a sea of smelly mud and dead beasts. The Garnet bridge was finally swept away, but the road is now passable for Landrovers and the like. The people from Gunnawarra, one of our neighbouring properties, went in yesterday and picked up our mail, which is now at another neighbour's place (Minnamoolka), so Ron will pick it up on his way. They also brought fifty loaves of bread in a trailer.

I have started mowing the lawn, which is over a foot high in places. Ron has mended the blades of the mower with the welder, but there's still a slight risk that it might suddenly fly out and cut me off at the ankles – an alarming thought. But the grass must be cut, otherwise we won't be able to see out of the windows.

Yesterday I took the kids for a walk in the horse paddock where we found some enormous *horse* mushrooms, which we carried reverently home and cooked in butter for supper. All the horses have swollen fetlocks from standing on wet ground and being unable to lie down to sleep.

Glen is sleeping in Sharon's bed while Ron is away, which saves a lot of traipsing around for meals, etc. Thank goodness she knows all about lighting plants, and can drive that big diesel truck. I would be all right on my own, but the responsibility of looking after a place like Meadowbank is a bit daunting.

28 March

The Hassalls were flown in last Tuesday by a bush pilot's little red Cessna, and we are now back to normal routine work as the weather has improved. In the splendid big mailbag that came next day was a letter from Merv asking me to go down and help him muster for three months. Apparently Twinkle is now shod and is the best of all the horses. It's tempting, but I can't just up sticks and go, abandoning everything here. Thursday was spent cleaning mouldy saddles and bridles and rusty bits, which is quite a job after all the damp weather – the green hairy mould spreads its insidious way over everything. I swear it will take over the world, or maybe just north Queensland.

On Saturday some neighbours picked up Ron, Glen, the kids and me to take us to Glen Eagle for the weekend to help salvage the Blakeneys' belongings. All the bush roads to Glen Eagle were almost totally washed out and the countryside was hardly recognisable, even to the people who knew it well. Glen and I with the kids were perched on top of an enormous load of swags, tents and bags of feed in a tiny 'Scout' (a small, International 4-wheel drive, 4-cylinder vehicle); on the way we passed a cow hanging in a tree about fifty feet above the river, its belly swollen and putrefying, its horns neatly stuck in the fork of a branch. On arrival at the station we found everything in a state of chaos. The temporary shed was standing on the base of the wrecked house, and there were piles of twisted roofing iron, upturned trees and yard posts scattered everywhere. Masses of people had turned up to help, and after a quick lunch Glen and I salvaged what we could while the men contrived a bathroom and hot water system out of a 44-gallon fuel drum with a fire below. I found a bottle of Old Spice aftershave, a pair of racing trousers wrapped round a branch, the lav cistern, a mop, a good girth and a dog chain; all these were put into the tent specially erected for salvaged goods. We set up camp but then it started to rain, so everyone huddled into the draughty shack and ate mounds of curry while the kids moaned and grizzled, doing nothing to ease the general atmosphere of sadness and shock.

When we went to bed the mozzies were so bad I had to cover everything bar my nose, which was consequently very red, itchy and swollen by morning. After breakfast I walked down to the Wild River (an apt name) to wash some girths and a couple of pairs of muddy trousers, and found some small crocodile tracks on the sandy bank.

The Meadowbank contingent left that evening after helping all we could, and had a pleasant but cold journey home in bright moonlight, sitting in the open back with our teeth chattering. Then

Meadowbank station, showing the room that Sharon and I shared.

Wabble Creek mustering camp: Ron, Sharon and Glen.

Me precariously perched on Bits, who sits down inconveniently at intervals.

Ron drenching Bits.

Dawn and I about to set off on a day's muster.

The night-mare, Sunny, at Meadowbank.

Pigeon, poddy calf, at Meadowbank.

Damona and I in Four Mile Dam.

Cows and calves at Moneymusk, Meadowbank.

Brahman bull in Meadowbank yards.

Mt Garnet picnic races.

Don't Talk, who won me 4 dollars.

With the horses at Scrub Dam: Harriet on Dandy and Glen on Imp.

Working horses in the yards at Meadowbank.

The remains of Glen Eagle homestead after the floods of March 1967.

Aftermath of the 1967 floods: a decomposing cow hangs 40 feet above the Herbert River.

314 Queen Street, Maryborough.

Jean Swinbourne and I, with Ian's speargun, in a Maryborough shop.

Ian, complete with the two fish that I caught.

yesterday Billy Blakeney, the son, arrived with a load of mud-stained clothes and linen that Glen had offered to wash in the machine. All yesterday and this morning we have been battling with coffee-coloured sheets and once beautiful damask tablecloths – a heartbreaking task because the stains would not come out completely. Still, at least they're as clean now as we can make them.

Later, feeling the need of a bit of light relief, we went out to muster a killer. I had a lovely ride on Shaneen, getting in eighty bullocks, including the killer, and yarding them just for practice. Actually it's almost impossible to bring one beast in on its own.

I hear the owners of Wiaruna (another neighbouring property) are negotiating to sell 500 Brahman cows to farmers in Ceylon in the near future. I wonder how that will work out. It sounds a bit like taking coals to Newcastle – but then beef production in Ceylon could well do with upgrading, if I remember rightly.

The bridge over the Wild River is now open again, and Sharon is coming back this week. She's had an extended holiday and missed all the action, though no doubt she will have some horrific stories of her own about the floods on the Tableland.

3 April

You will have to excuse this extraordinary scribble, but I am in bed recovering from some unknown fever or allergy. On Friday Glen and I went mustering, and the first time I got off Shaneen to spend a penny I found spots all over my legs. Then, coming home in the late evening, I had severe pins and needles in the hand holding the reins; otherwise I felt fine. Next day I was covered in a rash, rather like measles but more clearly defined, and yesterday my temperature was up to 102°F. Every finger swelled up and all my joints were tender; it was agony to move. Glen contacted VKA over the transceiver on the verandah outside the jillaroos' room. The Flying Doctor said that if the patient wasn't any better in the morning a plane would be sent out to take her to hospital, and on overhearing this I was so determined not to be flown out that I promptly began to feel better. Today I am much improved, though my fingers are still swollen. My temperature is almost normal, and I ate some breakfast, though I still feel very dopey. We have no idea what caused the trouble, unless it was riding through the young spear grass, which is about five feet high, and nibbling the tops occasionally. I don't see why that should affect me, though, as I've done it before.

The ride that day was the longest I've ever done: over forty miles,

107

with lots of trotting and cantering and chasing wayward cattle. Poor Shaneen's back legs kept folding up on the home straight, so I led her back through the horse paddock without realising I was sickening for something.

Saturday was April Fool's Day. While I was milking, Ron crept into the yard and set off a shatteringly noisy alarm clock just underneath the cow, on the far side from me. I don't know who jumped the furthest, Elsa or me, but no milk was spilt. The Boss had also put an alarm clock in a tin under my bed to wake me at midnight, but it failed to go off, thank goodness. Honestly, you can't trust anyone round here.

On Tuesday evening we killed the little fat cow Glen and I had brought in, and next day, while Ron and Glen butchered the beast, I did the wrapping and labelling of all the different cuts. Ron was acting the goat as usual, flipping bits of fat at anyone within range. Killing day is always certain to produce a good lunch; this time we had barbecued fillet, rump, skirt and liver. Lovely!

I shall stay on here until about November, then return to Maryborough for Christmas. This is a terrific place to work on, and I'm not all that keen on seeing the rest of Australia, even if I had time, as I doubt very much if anywhere could be better than Queensland. It suits me down to the ground, especially the northern and central parts.

Since being confined to barracks I've had streams of 'visitors' coming to say hello through the French window and piling books on the floor beside the trestle bed. I'm reading Mowgli again. The kids have been bringing me caterpillars and flowers, bless them, and now the Boss is about to report my condition to VKA, so I must sit up and listen. . .

Well, the doc. says I must have three phenobarbs a day for two days and cover myself with zinc cream. He also suggests leaving the spear grass to the cattle and not eating it myself; it could possibly be the cause of the fever, but he's never heard of it before. So, the plot thickens. I shan't be able to milk for a while because my hands are very stiff and sore and as puffy as pork sausages. Ron reckons I ought to haul myself out of my pit and go mustering with them tomorrow to find out once and for all if it is the spear grass or not – charming! They have decided I ought to wear Glen's egg basket as a hat to the picnic races at the end of the month. I might just do that; it's not a bad-looking article upside down.

9 April

The disease of unknown origin has more or less disappeared, leaving me with rheumaticky pains in the left thigh and knee joint, so that I am very lame and stiff in the early mornings. We start a concentrated three weeks of mustering before the races tomorrow, but I shan't nibble spear grass, just in case. On Monday afternoon I got up for smoko, ate a normal-sized supper and stayed up until bedtime. Mr Hassall tried to kid me into believing that I am known as Spear-grass Sue over the entire radio network of north-east Australia. I slept like a log that night and did light duties next day while the others went mustering, lucky things. They brought in a mob to go to the sale, which included a big speckle-faced yellow cow who charges everyone every time she is yarded, so she is a cert to go this time.

On Wednesday I staggered out and milked Wiggle for the first time, and also Gumnut, as Ron and Glen had left for the cattle sale and the Boss only likes milking Elsa, who has enormous tits. (Excuse my French, but that's an everyday word out here.) Wiggle, a Jersey-Friesian cross, is nice to milk but kicks like blazes, and her new calf Webster is a funny little object. Half-way through milking Gumnut I felt quite knocked up, so Mrs Hassall finished her off for me.

Angus, Sharon's ex-fiancé, left with the load of sale cattle and I had his room to clean out. *What* a mess it was in!

Everyone is race-minded now, and the talk is mostly about the forthcoming meeting on 29 April and 1 May. The Blakeneys (the Glen Eagle people whose house got washed away) are going to camp with us as obviously they haven't much left in the way of camping gear. They have a pleasant son called Billy, who is a good jockey and is riding in a few of the races. Can you imagine getting ready for a race ball in a tent with only a hand mirror and a carbide light, and nowhere to hang or iron clothes?

I have asked Jean Swinbourne to send up a tin of Baxters' Haggis from a large store in Maryborough, which I shall present to Mr Hassall for his birthday, as he shows so much interest in Scotland and the Loch Ness monster. By the way, on getting home I intend to *camp* on the banks of Loch Ness until I can take a photo of the wretched animal to send to these doubting Thomases. None of them will believe my confident declaration that there is a complete *herd* of Nessies lurking in the peaty depths of the loch.

You wouldn't think there had been 21 inches of rain last month. The grass is feet taller but is turning brown already, the ground is cracked and hard and things are wilting in the sun. Two of the paw-paw trees have died from saturation of their roots (foot-rot?) and the only things remaining to remind us of all that rain are the beautifully full dams plus the plague of mozzies. We spend most of our free time choking inside our nets, having sprayed the room thoroughly with DDT.

15 April

Sharon, Ron, Glen and I have been out mustering every day this week, from dawn until dusk. I seem to be back in good working condition at last, because I'm not at all tired or stiff – just ravenously hungry.

I rode Damona, on whom I have been practising mounting and dismounting 'bush-style' as opposed to the laborious Pony Club fashion normally used. The local way is much the same as the American cowboy method, except that you haul in the nearside rein, so that if the horse is thinking of bolting or bucking before you are properly on, he has to figure out how to free his head first.

Bits is coming back from Spring Creek tomorrow. I hope she's in foal – though that might make her meaner than ever. If she is, the result is going to be called Pieces, of course.

24 April

Al, the Hassalls' son, returned from his wandering life on the rodeo circuit, complete with three cracked ribs. He has no inclination to work on the place and tends to spend too much of what little money he wins on 'the grog'. He brought the mail from Mt Garnet with him, which included a letter for me from Merv. He told me that Don and Doug had turned up at Wendouree and were staying for six weeks, and that Doug 'incidentally, is sweet on you'. I had no idea Doug even remembered me, as it's a year since we met, and then only for three weeks, but Merv does say 'is' and not 'was'. I was in such a tizzy for the rest of the day that I forgot to salt the butter, put ginger on the junket instead of nutmeg, forgot we had put bulls out in a certain paddock only the day before, and was thoroughly absent-minded for hours. I might see Don and Doug at Home Hill and Mareeba rodeos (May and July), and Ron and Glen are asking them to come here for a holiday after Mareeba – yippee!

Now, back to business. On Wednesday I caught Bits, and Glen roughed her for me in the yard before I was brave enough to mount.

She was very docile, however, except for trying to bite me on the bottom, which she usually does anyway. We mustered the bull paddock; Glen reckoned I was still in a bemused state after reading Merv's letter, so she kindly sent me round the fence in case I got bushed.

Thursday I spent cooking for the races, making piles of cheese biscuits and bottling rosellas from the garden (not parrots, but peculiar acid fruits that look like small, spiky, red octopuses growing on a bush).

Friday was misty and cold. I milked Elsa under a dripping tree, then went mending fences with the Boss in the Landrover; I had to drive while Mr Hassall walked ahead and made a vague path for me to follow over the basalt boulders. The fence was in a mess, just over the creek where the floods had smashed through. The smell of something dead kept us company, while nasty wetting little April showers fell throughout the day; one minute we were cowering under feed-bags beneath a tree, the next we were steaming like wet dogs.

That night as Sharon and I were getting undressed – my frilly knickers were around my ankles – the door was rudely banged upon and opened, and in dead silence in walked a *goat*! Loud laughter followed on its heels, and the entire household crowded into our bedroom. The billy jumped onto Sharon's bed then stood on my toe before Ron carried him out again. We stood completely stunned and unable to utter a word until the extraordinary procession had gone, closing the door behind it. Later, when we'd recovered somewhat, we learnt that Boomerang Station had lost forty goats some time ago. Apparently they usually drift east, and this one happened to stop here; unfortunately for him, because we had half of him for supper and the other half is in the freezer ready for the races.

Yesterday afternoon Glen, Al and I mustered horses and brought the whole lot in to be drenched for worms today. I was sent off with Al, me riding Shaneen, but we only found five, while Glen got the other thirty or so. How Al rides with cracked ribs is beyond me, but he seems okay. He is a man of few words. Shaneen pulled badly and had to be brought round in a complete circle to slow down at one point. She's losing condition, so will have to be turned out for a spell.

Today is my day off, and the others have been drenching the horses. Bits nearly knocked Ron down and had to have a nose twitch put on; trust her to be the only awkward one.

30 April, The Race Course, Mt Garnet

The ladies of Meadowbank spent a frantic few days before the races making hundreds of biscuits and loads of white sauce to be frozen and used for thickening soups, stews, etc.; then on Thursday Glen and I and the kids left in the VW for Mt Garnet. On the road we met the Boss and Ron on their way to a sale in the Landrover, so we swapped vehicles and I drove the rest of the way in, getting stuck behind a horrible road-grader that spat mud everywhere. Glen persuaded me to drive into the town and get my licence changed; all the cop made me do was park, reverse out, do a U-turn and a hill start, for which he charged me 4 dollars. To celebrate I did all the driving that day, even in the town to collect food, food and more food, which was taken back to camp in relays. Our canvas 'village' was all ready: one enormous tucker tent containing two long tables, a fridge, washing-up table, shelves, meat safe, and various cupboards and crates for storage; and four other tents for sleeping in. That evening I made a huge pan of fried rice over the big campfire, adding about three pounds of steak grilled in strips, after which we went to see some slides of the floods and the damage to the Wild River bridge – all quite horrifying. Glen and I had lovely hot showers and went to bed quite early on our double-decker bunks; I had the top one.

All the time we've been here it has been freezing cold, with a raging wintry wind and nasty showers; it's a great shame, as there would have been an even bigger crowd if the weather had been good. Even so, there's a reasonable turnout. On the Friday morning I awoke to what sounded like a big cat purring in my right ear, which turned out to be the strumming of a distant guitar from another camp after an all-night party. The Hassalls and Sharon arrived that day, and in the evening she and I were taken into town by Ray Cowan, a pleasant young grazier friend of Ron's. The town was all done up with lights and merry-go-rounds, hundreds of people and almost as many dogs. One of the latter came into the dance, and every time his mistress got up he followed at her heels, dogging her footsteps quite literally until he found himself doing a passable waltz. He lost her once, and went galloping round the floor sniffing at people's hems until he found her again.

I danced with Ray, Ron and Al. I can't help liking Al in a strange way. He's a bit moody, with a fair old chip on his shoulder, but is appealing for all that. Ray took Sharon and me home, waited while we changed into jeans and layers of jerseys, then took us to find a party where there was lots of singing and fooling around. Later we discovered a younger set who were sitting around a huge bonfire, so we didn't hit the hay until 5 a.m.

Two hours later we were up, to spend the morning preparing mounds of cold food and getting ready for the races, which were held in the afternoon. I backed a nice-looking chestnut called Don't Talk from Spring Creek and he went on to win his race, gaining me all of $4 – in fact paying for the driving licence. He thought nothing of his blue winner's ribbon, however, and plunged round and round it trying to throw it off. Billy Blakeney from Glen Eagle had bad luck with his first two rides, but he won the last couple splendidly, sweeping up from last to win.

Ray took me to the Race Ball that night, but it was terribly crowded, the band was lousy and the floor had large cracks in it, so we left before supper and I was in bed by 1.15 a.m.

A lot of dubious things go on in various tents – not surprisingly – which keeps us entertained. We didn't see the two girls in the next tent all day until they emerged at dusk, bleary-eyed, to collect some roast chook, returning to muffled coughs and chuckles. Sharon is dying to know who their men-friends are! Today she and I are doing absolutely nothing but catching up on lost sleep, washing our hair and preparing a party for Al's twenty-first. He, however, has disappeared with some of his mates in town, and no one knows if he will turn up in time for his family celebrations. He's quite likely not to.

Peter Darvall, the vet, is here with his wife, new baby, two dogs and a cat, all in the smallest tent you've ever seen. They are a lovely family.

Monday   Well, what a terrific time was had last night. Ray took Sharon and me to the dance and we danced every number between 8.30 p.m. and 2 a.m., all with different chaps. I wore my navy and white short dress and white bolero, which was quite effective, apparently, and we returned to camp footsore and lame, to be swept off to a party by Al and a bloke called Ross, an English pom. Al, for all his being a bit of a black sheep, can be very charming when he wants to be, and we all had great fun at a fireside shindig that eventually broke up at dawn. I managed to sleep soundly until 8.30 and don't feel too bad today – yet.

8 May, back at Meadowbank

We are back in routine now after the races, but I'd have that week all over again tomorrow if it were possible. The drinks party for Al was a success – he arrived back in time for it, thank goodness – and the dance that night was very enjoyable too. Just when Sharon and I were about to go to bed a car drew up outside our tent; Ross called out wanting to take us to a party, to which we agreed with alacrity. Al was there, slightly drink-taken and in sentimental mood. Sharon took her record-player to the fireside and we listened to records until dawn, when the batteries began to fade and the party reluctantly broke up. Next day I won another $3 on a very exciting race where I was egging on the wrong horse. He was third, while my proper horse was way out in front! Ray let us girls go into town in his Nissan, a powerful brute, which Sharon drove while I sat petrified, but we got there uneventfully to enjoy the dance, each having hundreds of different partners. In the gypsy tap Al kept returning to me after every change of partners; I don't quite know what to make of him yet.

Tuesday was packing-up day – what a job! And the cold howling wind didn't help matters either. I ended up going home with Al in his red Ford Falcon and was pleasantly surprised by his careful, steady way of driving; he doesn't race along like most others of his age. He left Meadowbank next day to work for a while on another station and will return to pick up Sharon and me for Home Hill rodeo, where I hope to see Doug Spann.

Glen has gone to stay with her sister Rome in New South Wales, taking the kids with her, so the house is terribly quiet. Milking is rather a shambles, too; the Boss does Elsa while I struggle with Wiggle and Mrs Hassall comes down to do Gumnut. Wiggle hung up on me the first day just before the cream, as presumably I was much slower than Glen at milking.

Yesterday we mustered Billygoat Paddock to get some spayed cows in for sale. I rode Bits, who must definitely be in foal as she is nastier than ever. Sharon had to hold her head while I saddled up, or else I would be minus a couple of fingers and some denimed bottom. Soda, the kitten, came down to the yards and jumped up the rails onto the post beside Bits, who flung up her head like a great stallion, her nostrils red and flaring, and trembled all over with horror. Then while crossing a creek she rooted all the way up the far bank and did a circular dance on top with her head down between her fetlocks; all quite comfortable, but it gives one an idea of what she could do if she put her mind to it.

So now the jillaroos are back to weeding, sweeping, cooking and milking. The next excitement will be Home Hill rodeo on 27 May.

13 May

Maybe it was dengue I had, as you suggest; I should have thought of that before with all these mosquitoes about, but then one doesn't get spots with dengue, does one? My fingers are still weak and sore and the twice-twisted knee is painful, so perhaps your diagnosis is right.*

Milking Wiggle this week has improved my technique somewhat; I seem to get as much milk as Glen now, but without the 3 inches of foam on top. Mind you, she's been milking since she was knee-high to a grasshopper – well over 7000 times to date, so she should be pretty good at it by now.

Tuesday morning was spent digging khaki weed out of the middle of the road in the horse paddock. I was wearing my swimsuit and of course a rare visitor came through in his Landrover, peering out with raised eyebrows, hoping to be asked to stay for lunch. Ah well, at least my back is nicely brown now.

On Thursday Ron and Sharon went out road-making while I started to paint some signs – two with GATE on them and one with MT GARNET; then I painted the kitchen table pale green. I had avocados with cream dressing for lunch – better than anything The Ritz could possibly provide! On Friday it was my turn to labour for Ron on the road-mending job. We set out after breakfast in the old Morris truck, made a new stretch of road of about one and a half miles, then rejoined the old road and started digging up the wretched basalt rocks that were sticking up in the tracks. Ron did most of the digging, but I had to fill in the holes after him and cut out roots with the mattock, becoming weaker and crankier as the day wore on, and finishing up with a blister on one toe. It drizzled on and off all day. At one point Ron lit the grass to see if it would burn, and only just got the truck out of the way in time, as the fire spread rapidly, despite the damp atmosphere. I drove part of the way home, learning to 'double de-clutch' the noisy old Morris.

Mrs Hassall gave me an old R.M. Williams 'Longhorn' jacket (worn by stockmen, etc.) that Al found on the road, so I've been patching it up and now it's wearable at least. My signs are finished and were erected today; we're hoping no one will come along and shoot at them. It seems to be the fashion around here – every signboard on the main road is full of rifle bullet holes.

The weather is still cold and blustery. I sleep with four blankets and I've made myself a pair of the most abominable slippers you've ever seen. One is pink, the other a faded purple, but at least they're warm.

Mr Hassall is breaking in a filly just now, which is most interesting to watch. He rode her for the first time today and reckons I ought to try my hand at breaking. I'd love to, but haven't the time nor the courage to start, especially with someone else's animals.

Wednesday morning the mailman rang to say he'll leave our stuff

---

* It was in fact the mosquito-borne Ross River Fever that I had.

at the turn-off, so I was sent out in the Landrover, complete with an empty gas cylinder, a drum of water and a hoe (to attend to the bamboo shrub growing near the main road), and drove the 17 miles to collect the mail. Ron and Sharon were working on the Road Improvement Scheme, and I had to avoid a whole lot of gaping holes and unearthed stones. On the return trip I was forced to squeeze the 'rover between the Morris and a tree, with loud cries of 'Whoa!' from Ron, which startled me somewhat and amused him greatly.

That night Al rang up and said he had been kicked on the wrist by a horse and that the muster had been held up a little, so he didn't know if he'd be able to get here in time to take us to Home Hill, though he'd try. As yet he hasn't rung again, and there are only four more days to go. Perhaps I shall have to wait until Mareeba rodeo in July before seeing Doug again, but as we haven't met for over a year anyway it won't make much difference.

Thursday was a rainy, miserable morning, but I went out road-making again and worked flat out all day, first of all loading the truck with basalt sand from the creek bed. (The Morris almost begged as she filled up with the heavy sand.) I unloaded it all into the holes that Ron had made while digging up rocks; then it was lunchtime and I had an awful job getting the fire to burn in the drizzle. Later Ron blew up some stumps with gelignite, having made sure I was safe behind a distant tree, then at five o'clock we loaded the truck again and blinded a good long stretch of stony road. Macalpine's fusiliers have nothing on the Meadowbank jillaroos when it comes to road-building!

Zipper, the new filly the Boss has been breaking in, is a beautiful animal, a pale creamy colour all over with enormous dark eyes and film-star lashes. Poor thing, she still isn't used to having a weight on her back, and when Mr Hassall put her through the spring she plunged and tripped over the rocks trying to keep her balance, at the same time doing her best to spit out the bit. He's very pleased with her, though. Riding youngsters gives him a terrific boost, as he reckons he can't be all that old if he can still break-in successfully. He wants me to have a ride on her, but I think I'll wait until she understands her bridle a bit more.

Did you hear about Elvis getting married? Nobody told me until about a week later, when I happened to see his photo in the *Cairns Post*. Good luck to the wretched fellow. He hasn't had a very happy life, for all his riches.

30 May

Al arrived back in the middle of Friday night, and on Saturday morning the three of us set off for Home Hill some 300 miles away, leaving at 8.30 and crossing the Kirrimer Range to join the highway at Cardwell. Just past Townsville a battered Falcon roared past, hooting. Al drew in and stopped, then he and the boys from the other car had a drink together while Sharon and I sat fuming with impatience.

We got to our destination just before sundown, and the first person we saw was Don Hafemeister sitting in someone else's truck. My heart sank, thinking Doug hadn't come and Don replied to my question, dead pan: 'Nope, he's gone back to Brisbane – givin' up rodeos'.

'Oh', I said, concealing (I hope) my disappointment.

'But he decided to come up here first!'

Rotten beggar! I'd forgotten what a tease Don is. Doug was in fact not far away, grinning at Don's little joke. It was just great to see him again, and the feeling seemed to be mutual.

Sharon took me to meet her cousins who live on a cane farm, where we had supper, changed for the dance and went straight back into town again. I had a fabulous time, dancing mostly with Doug, but also taught Don my version of the jive, which he picked up quickly, when he wasn't clowning about. However, his various broken bones started to play him up so we had to slow down. The dance ended at midnight and I went back with Doug in his Ford 250 truck while Sharon and Don followed on behind in Al's car. The roads through the cane all look the same, which proved troublesome, but after much reversing and a lot of wisecracks and laughter we got back to the cousins' farm in convoy. We sat and talked for hours about cattle, horses, rodeos and other bush subjects, and the time just flew by. Doug says when he gives up rodeoing he may take a permanent job with someone like Merv and help him run the place – but somehow I can't see him ever giving up his wandering way of life to be tied down to one small part of the country.

The next day was the rodeo itself, a most exciting spectacle. Don got caught up on his bull and hit his knee resoundingly against the rails, so didn't do all that well in the bareback, though he came third. Doug was second in the bull ride, narrowly beaten by half a mark, and was also second in the bulldogging, helped by his delightful little pony Twinkle. She's looking very well, but still goes off her feed before a show. The bulls used for the bucking events were huge black Brahmans, one of which had disembowelled a horse the previous week: they were sinister-looking beasts with evil tempers. Al won the wild-horse race and the show finished at five when Sharon and I returned to the cane farm for tea. Doug came out later to see me for a while, then he and Don had to leave that night. They have a job breaking horses for six weeks until Mareeba rodeo. Roll on that day!

Next morning Al took Sharon and me back to Meadowbank, stopping at Townsville for three hours to get the car some new 'shockies' and a radiator hose. Six hundred miles may seem a long way just to watch some bucking stock, but to Sharon and me it is well worth it and we can't wait for Mareeba.

Last Monday Shaneen took me for a lovely ride. On Tuesday I seemed to spend all morning sitting in the big truck while Ron grovelled underneath and bled the air out of the brakes and clutch; I jumped up and down on the pedals at various intervals, which presumably helped. That evening a Bush Brother arrived with his dog, having got bushed, so he stayed the night and kept us amused at supper with his entertaining stories of religion in the bush. Sharon and I had finished off the road improvement scheme with a truck-load of sand that afternoon, and put a log across the old road to warn Ron to use the new one on his way back with a load of lucerne. Brother Bill came along, saw the log, decided not to drive round it in typical Aussie fashion, but got out and removed the offending article altogether; he thought it was blocking the right road. This caused great chaos and much merriment later when poor Ron trundled home on the bumpy, twisting old road with his top-heavy load.

On Wednesday the two jillaroos went mustering bulls with Mrs Hassall – definitely a 'ladies' day'. I rode Bits, who shied badly three times and nearly threw me. Then she rooted once, almost pulling me over her stubborn head, and on the way home she took off like the clappers and galloped for about half a mile, clattering over stones and slipping on the seepage outside the dunny before stopping at the yards, not even panting. I was, I might add.

Ron and I did a bit more to the road the next day, and just when he had dug up a patch of big rocks, three army jeeps came tearing along in one hell of a hurry. They ploughed straight through the holes with much yelling of oaths, leaving us in a cloud of dust looking at each other in amazement. We never found out what it was all about, but presumably the army lads were on manoeuvres or even a treasure hunt. Anyway, apparently they got bushed near the homestead, ending up in the boggy part of the milkers' paddock with three cows staring at them incredulously.

5 June

I have recovered from Home Hill, more or less, but still tend to be absent-minded and do crazy things. Maybe I *am* crazy!

I went for the horses on Sunny the other day, and she proved once and for all that she *is* a night-mare. First of all I scrambled on bareback and rode as far as the wire gate, where she dug in her heels, snorted and behaved in a thoroughly silly way, as only she can. I slipped off and led her through. The horses were up the hill and Sunny wanted to canter down after them as they crashed over the basalt rocks. When I wouldn't let her she tried to grab the bit, her chin almost on her sweating chest, and of course, not being able to see where she was going, nearly did a series of somersaults into the yards. A horse of little brain, I fear.

I had the afternoon off and started to make a white brocade dress for Mareeba rodeo dance and the Oak Park picnic races in August. Then some people called Wilson came and stayed the night, bringing with them an unbroken filly, which Ron and Glen bought for the kids. She is a smallish chestnut pony called Imp, and should turn out to be quite good, as she has been handled since she was a foal. It never ceases to surprise me, though, that horses out here are not even backed until they are 4 or 5 years old – by which time one would have thought they'd have learnt all sorts of tricks.

On Sunday we all left early to spend the day at Whitewater, where we were a couple of months ago. This time the garden was green and prolific, the weather cloudy and temperature 70°F, whereas last time we sweltered at 103° with a hot dry wind. The men went off to a meeting, leaving all the women and some noisy kids in the house. Sharon and I became somewhat sated with the 'galah session' and went for a couple of walks up the creek to fill in the time until the men returned; a houseful of women really is the end. Still, there was a lovely lunch of cold turkey, pineapple salad and bread and butter.

Today the Boss and Ron left for the Mareeba Cattle Production School, which lasts three days; then Glen flies up from her stay in New South Wales on Thursday with Tammy and Dave, so we shall soon have a full house again. Once everyone has returned we start a concentrated fortnight or so of weaning and preg-testing, on those horribly fresh fat horses – help!

Al was given a young cattle-dog the other day and has named her Susie – after me, so he says. That's a dog, a dinghy and a big fat cow all using my name. Didn't someone once say that imitation is the highest form of flattery?

12 June

On Tuesday night 97 points of rain fell, followed by another cold dripping morning, but as it was my turn for a lie-in until eight o'clock I snuggled down in bed and revelled in the thought of poor Sharon milking in the wet. Extraordinary weather, really. Brisbane is having its worst floods for a century, with much damage, and Surfers Paradise is feet under water. Meadowbank has had almost three inches during this week; the first winter rain for twenty years. That's because I'm here!

Last Saturday Sharon and I were sent out to muster horses. On returning to the house we were told that we were going to the next-door station, Boomerang, for the night; we had ten minutes to wash our hands, pack swags, change and pile into the Landrover. Ron had to go into Garnet to his Lodge practice, all terribly private and masculine, and we mere women spent the night with Ellen Honnery, listening to Ranch Club on the wireless, bathing the kids and generally having an easy time. I spent a restless night balanced on the edge of a double bed hanging onto the sheet every time Sharon turned over. She's not the best person to share a bed with if one wants any covers left. The men crept in noisily at 3 a.m. and we went home in the morning.

A couple called Wescott came for dinner this week, and I showed my latest slides on the screen. When we came to one of Doug and me that Don took at Home Hill, the Boss said, 'Now that's what I call *real* friendly!' and everybody roared with laughter, to my acute embarrassment.

Today Sharon and I cleaned the VW fastback, successfully emptying the water tank in the process, then we greased our saddles, using kerosene in the beef dripping to stop cockroaches eating the leather. The beastly things have chewed a hole in my lovely pink wool jumper and several other natural fibre clothes. They swarm out at night; you can hear them rustling about the place.

19 June

Early on Thursday morning we set out to muster Top and Bottom Brumby paddocks and Bits, the bitch, threw me in the creek. We crossed it too quickly and she just ducked her head down, did two almighty bucks and off I came, thud onto my back. It's a long way to the ground from her vast height, and I got spear-grass seeds and black soil all down my pants. However, I climbed to my feet none the worse, save for a few bruises, to see Bits glowering sardonically at me. Ron, Glen and Sharon witnessed the incident and assured me it wasn't just a pigroot but a 'fair dinkum buck', so I didn't feel quite so mortified – though they might well have been trying to be kind. Bits was quite good for the rest of the day, and we spent some time galloping and stopping when *I* wanted to and not when *she* wanted to. We had lunch at the dam – the usual beef sandwiches and tea out of quart pots – and then I was left with 400 head on the flat. The Boss came along and said, 'I wish your parents could see Sue the jillaroo now – all alone with a big mob'. I'm glad you didn't see me a couple of hours earlier in the creek with my legs in the air!

Next day was 'Black Friday', when everything went wrong. We preg-tested all day, starting at dawn and working steadily through the herd. The Boss was de-horning a heifer when she suddenly tossed her head and sent the tippers crashing against his mouth, gashing his lip and chin, which bled profusely. The poor bloke went back to the house wondering if he'd lost any teeth. Soon after, Sharon left the field when an irate high-grade Brahman cow scraped her leg against the top rail, bruising it quite badly as she was trying to clamber over the rails out of the way. Ron, Glen and I carried on with the work until it became too dark to read the ear tags easily, then Glen took Davey on Sixpence to put the dry cows into the nearest paddock. Unfortunately they came out of the yard at a run, and Sixpence became overexcited and bolted, bouncing 4-year-old Dave off after a couple of bounds. Ron said 'Christ!' and dropped everything to run out of the yards and pick him up. He wasn't hurt at all, thank the Lord, but of course screamed long and loud for some time with shock, poor kid. It was a subdued household that night, for a rare change, all of us quite convinced that accidents definitely come in threes.

Next day was lovely by comparison, and everything went smoothly. Having finished the testing early on we took the pregnant cows out, me on Dawn, who was very fat and fresh but well-mannered nevertheless. The cattle thundered out – a splendid sight, almost 400 prime cows of all different shades of red and gold, white and brindle. After depositing them on the water we went off to muster another paddock to find 240 head for pre-testing. This was done after lunch, when Al and the Gunnawarra jillaroo, Harriet Scott, turned up for the weekend. She is a nice girl with a pleasant sense of fun; her parents call their house 'Wha Hae' because of their surname. What would Wallace have said about that?

The climate has gone quite crazy again – it is now pouring with rain. This morning I awoke with rain on my face so had to reach through the mozzy net and close the French window. Then Hennessy, who was on my bed, wanted to go out, in his typically cat-like infuriating way, so it was a disturbed dawn. However I had ten blissful extra minutes in bed when Sharon had to get out and milk. The morning was spent making peanut biscuits and pikelets, all of which disappeared at smoko. Mrs Hassall says she dreads my cooking week because everything goes as soon as it is made instead of lasting a couple of days. We heard on the radio that there are more floods on the coasts, and the cane-crushing season has come to a halt for the first time for years.

26 June

Wednesday was another cold and miserable morning spent in the house cooking, washing and cleaning. Thursday wasn't much better, except that I managed to get into the garden and prune all the dead prickles off the cumquat and lemon trees; then in the afternoon Glen, Sharon and I chased the 100 head of weaners into the crush to read their ear tags and weigh them. One little bull knocked a couple of his mates down and used them as a ramp to run up and jump over the top of the crush to the freedom of the big yard.

I plucked up enough courage to ride Bits to muster the pregnant cows out of Bowerbird Paddock, and she was quite horrid while saddling up, stamping her feet, refusing to stand still and nipping my backside. Once on board, however, I had little trouble with her, though admittedly I hung onto the monkey grip while crossing the creeks.

Today saw my last ride on Shaneen. She has been turned out for a spell now, because she falls back a lot during the winter, although she's thin and scraggy at the best of times. First, though, she and I helped the Boss take a batch of weaners out of the yards and back to a paddock. They were as good as gold, which goes to show that a week being fed in the yards makes all the difference to their behaviour on release. For lunch I made a big pot of Scotch broth for Ron and Glen to come back to after their day's fencing. They were delighted and loved it despite the fact that I had to use onions instead of leeks. Sharon and Mrs Hassall are away for a few days.

I'm very tired tonight, so must sleep now. There are seven blankets and a cover on the bed, I wear socks as well and Hennessy sleeps on my toes, like a tabby rug with lumps in it. The cold at night is really penetrating, not unlike that experienced in the desert in its intensity.

3 July

Glen, Ron and I went to Lochlea station for a couple of nights this week. The house was the coldest I have *ever* been in, with a stone floor, draughts whistling in everywhere and a tiny stove to keep us warm. Of course it was the coldest night so far, with a slight frost in the morning. Next day we all went up to The Lynd near by, where Ron was going to teach the people how to pregnancy test. I have tried a couple of cows myself, one three months in calf and the other dry, to feel the difference in the ovaries. ('Greensleeves', Ron aptly calls himself during the process.) The young couple on The Lynd, Norm and Vicki Kippen, are very nice indeed. They gave us a lovely lunch of prawns from Karumba, in the Gulf, lots of salads, and tarts with cream; then we spent the afternoon lying out on the grass looking at slides. Towards dusk we helped them kill a beast for meat. I wish I had taken my camera, because there would have been a dramatic shot of the dead beast's head lying in a pool of scarlet blood with the setting sun making long shadows of its horns. Rather morbid perhaps, but eye-catching and certainly colourful. We returned to Lochlea for supper of roast pork – the first I've had since leaving home, except for a greasy chop at Innisfail on the way out here – and returned home next morning.

Friday and Saturday were spent mustering with Glen and Sharon, doing Moneymusk Paddock for a stranger bull that was put back in the neighbour's paddock, then Billygoat Paddock to fetch in and kill a big fat bullock for Peter the vet to collect later for his deep-freeze. In the evening the Boss was thrown three times from the little pony Imp, which Glen has been breaking in for the kids. He refused to let her be the first to ride the filly, so he got on, and for the first couple of days Imp behaved very well; then as I said, on Saturday the Boss was thrown very determinedly three times. Fortunately he wasn't hurt, but it does seem a bit much to expect a temperamental 4-year-old pony to become quiet and well-broken enough to be a reliable child's pony. Somehow I can't see Imp ever settling down properly.

Today is mine, to catch up with sewing, letters, scrapbook and other odds and ends. Mareeba rodeo is on the 15th, then Oak Park races next month; they are having a Roaring Twenties night, so have you any ideas as to what I can wear without having to make a completely new dress? I thought of lots of feathers, which would entail having to catch an emu or something, but Mrs Hassall says she is going to kill a rooster soon, so perhaps there'll be some scope there.

12 *July*

Thank you for your letter, headed 'My Flipping Jillaroo'. Well, my dear Mother, of *course* I'd stick on the horses more *if I could*; I don't bale out on purpose! After all, one could hardly call the St Leonards school horses buckjumpers, by any stretch of the imagination, and as they are the only horses I have ridden before, what can you expect? I am learning but the mongrel animals always take me by surprise; naturally it's the best way to get rid of an unwanted rider. Yes, I have thought a lot about returning, and will almost definitely be leaving next year. There's more than $1200 in the bank to pay my fare out again should I find it impossible to adapt to the narrow confines of Scotland once more. The life out here is my idea of the ideal way to live (apart from the eternal early rising before sun-up), so it seems I might well fret for the wide open, lightly timbered spaces of inland Queensland.

Ron is now riding the new pony, who has become a really nasty little character. He hasn't been thrown yet, but Imp bucked and nearly fell down once and Ron had to push her back up by jamming his leg out straight against a post. Today he took her out mustering for the first time, but she didn't go at all well, bucking and rearing. She may have to be scrapped as a liability.

Thursday I spent paddling in sump-oil and thoroughly enjoying myself, greasing the forage harvester, big rake, the plough and grubber. Not that these machines ever seem to be used; they're just there in case they are needed, and have to be maintained.

Saturday was a day off and I tried on my Roaring Twenties outfit: a pink silk dress belted round the hips and worn over Mrs Hassall's pale orange skirt, which comes to just below my knees. I've made a headband with some of the deceased rooster's droopy feathers, dyed lilac, sticking out of the band around one ear, and have put straps on my white heeled shoes. I tried to flatten my bosom with a bandage, without much success.

It was Glen's birthday on Sunday, and I gave her two paperbacks: *Seal Morning* by Rowena Farr and *My Family and Other Animals* by Gerald Durrell. She seemed very pleased with both; the Durrell book is hilarious, I think. Later I ran the horses in on foot, using Glen's stockwhip, and succeeded where everyone thought it hopeless – pure fluke, of course! Today is Tammy's third birthday. I gave her a small plastic engine that issues a series of rapid squeaks when pushed. She was delighted with it and has taken it to bed with her.

18 July

I am absolutely dead tonight, suffering from a bad attack of the Mareeba rodeo hangovers.

Mr Wellman the mailman came on Thursday instead of Wednesday, reversing his 500-mile journey so that he called here last to pick up Sharon and me and take us and our swags into town. A few miles before we reached Garnet he ran out of petrol, so had to fill up from a drum on the back. As Sharon held the pipe into the tank, the blowback of air blew a great gush of fuel all over her and she stank of petrol from then on into town, poor girl. We were dumped in Garnet where Al was waiting to take us on to Yungaburra, on the Tableland, where Sharon lives, before going on to the showground himself. I was glad to get to bed and slept soundly, waking on Friday morning to find the Tableland drizzle descending in cold, grey shrouds, reminiscent of upcountry Ceylon in the monsoon. Sharon introduced me to her three horses and foal, then we packed up the car and drove to Atherton, reaching there about ten o'clock. I bravely went and had a tetanus injection; there are three more doses still lurking in the fridge, biding their time. Eventually we hit Mareeba and set up camp next to the toilet block – *not* my suggestion, but no doubt it was 'convenient'.

We spent the afternoon walking around watching all the preparations going on, the roughriders arriving with their ponies and trucks and the bucking stock being drafted in the yards behind the arena. In the evening there was a big parade in the main street – lots of beautiful horses, fancy dress, decorated cars, tractors and bicycles plus the town band perched on a truck playing madly and a pipe band which, not unnaturally, alarmed several of the horses. Some of the onlookers nearly bolted too – but the Caledonian Societies are very strong all over Australia. It was bitterly cold standing out in the street; Sharon and I shivered non-stop until we left and went in to a dance to warm up. Al weaved his way into the hall with his face all bruised and purple and swollen to twice its normal size – he had got into a drunken brawl with a couple of Aborigines, silly lad.

Back at the grounds after the dance, I saw Doug's truck had arrived. We didn't meet until next morning before the show, however, when there wasn't much time to chat and he was a bit tensed up anyway. He didn't do anything spectacular: got thrown off his bull before time but scored quite well in the bareback. It was terribly cold all day, the crowd was muffled up and shivering in the icy winter wind. Don's horse broke its leg in the saddle-bronc event – the sharp noise was clearly audible – and went bucking on round the arena on three legs before the pick-up men raced in and removed it. This unpleasant incident put a bit of a dampener on the show, and when lunchtime arrived Sharon dragged me off to have a go on the 'Tilt-a-wheel' to try and cheer us up. Never will I do that again! It goes round and round and up and down while the seats go in the opposite direction at great speed, making one feel airsick, seasick and just plain sick, as if one's neck is being broken very slowly.

The dance after the rodeo was much more fun; Doug took me to it and we had a lovely time. Even in the smoochy dances, when everyone was becoming sentimental, we were still chatting away about cattle and horses. He left me for a while, during which time a little Italian descended, told me I looked 'Boooootiful' and offered me some chewing gum. Trying not to laugh, I had great difficulty in keeping him at arm's length for two whole dances, having no idea that Doug was up in the balcony with some of his mates having a good chuckle at my discomfort. However, he then came to my rescue and we danced together after that.

I slept like a tired dog in my swag that night and got up for the second day of action. Doug rode his bull successfully in the next round, but as he jumped off at the end of the eight seconds, landing on his feet in good order, he pulled a couple of tendons in a leg. The silly fellow then went in for the bulldogging, where he reached the finals, but made his leg so much worse that by the time night came he could hardly move, let alone dance. Instead we sat and talked well on into the night, arriving at the mutual conclusion that much as I adore him (and vice versa), it wouldn't be sensible to let things get too serious. We don't have enough common ground and, besides, rodeos are not much of a life for a girl, with the constant travelling and injury and lack of money. He intends to go on for a year or so yet, so he says, but somehow I can't see him ever retiring from 'the circuit'.

Don, on the other hand, has decided to give it up and settle down to a station-hand job. He intended selling his spurs, bareback rigging and bull ropes after Mareeba; all very sad for him, but he has enough sense to pull out at the right time before he's killed or permanently crippled. Both he and Doug have already broken nearly every bone in their bodies. There was one horrible accident on Sunday; a rough-rider fell off his horse, landing flat on his back and somehow jarring his head on the end of his spine, which caused him to have a fit. He shook all over and went stiff, turning a hideous purple colour while blood streamed from his nose and his eyes rolled upwards. Then he went limp. Doug ran out to him and thought he was dead, as he had stopped breathing, but the ambulance men got him going again and he came to in the hospital a few hours later feeling none the worse.

Yesterday Sharon and I reluctantly said goodbye to Don and Doug and came back to Gunnawarra with Al, where Ron picked us up to take us home: two wrecks of girls dead beat on our feet, though thoroughly happy in a sad sort of way. I doubt if I'll ever see Dougie again.

31 July

Having taken a few days to recover from Mareeba rodeo, Sharon and I went with Glen to fetch in a killer, which was shot on Thursday morning. The meat hung in the shed until Saturday – just as well there are no blowflies around – while the rest of Thursday was spent cleaning the ceilings and picking guavas for bottling. Everyone except Mrs Hassall went fencing. She was skinning the killer's feet to make brawn when I heard her call out; she had slashed a big vein under her ankle with the butcher's knife and dark red blood was flowing out, much to the interest of the cats. I helped her back to the house, bound up the cut and put her to bed with a cup of tea and the medical book. The wound seems to be healing quite well now, thank goodness.

At the weekend we cut and wrapped up the joints and minced the scraps – I wouldn't mind being a butcher, if I were desperate for a job – and the ribs were roasted over a bark fire for lunch. At three in the afternoon Mrs Hassall took her jillaroos 100 miles to Ravenshoe to see *Dr Zhivago*, a strange film, depressing but absorbing, with excellent acting and beautiful scenery. We spent the night in the pub on nice comfy beds, had steak and eggs for breakfast, followed by a milkshake in Mt Garnet, and came home. On the way Sharon and I decided to go gem-hunting in the Anakie gemfields (100 miles east of Alpha) when we leave Meadowbank, or at least after Christmas in the heat of summer, when there won't be many folk there. It's a terrific idea. Who knows, we might make our fortunes out of sapphires and rubies!

I spent the whole of Monday morning hammering springy steel wire straight to make fence droppers – 160 of them – gaining sore hands in the process, while Ron and Glen left for a fishing trip off Cairns. Next day I beat out eighty more droppers and bottled further supplies of guavas. We had some quite heavy rain in the afternoon, which prompted Sharon to start singing Slim Dusty's 'When the Rain Tumbles Down in July' – an event unheard of in these parts.

Next day was spent pruning a cumquat tree in the chook yard and cursing the prickles, while the hens threw hysterical fits every time I moved. Al came through on his way to Hughenden rodeo, lucky beggar! I was sorely tempted to go too, but went wood collecting with the Boss instead, having asked Al to give my love to Doug. Ron and Glen returned later from an enjoyable boat trip, but the car was full of glass as their windscreen shattered just outside Garnet.

Yesterday Glen and I went out for a day's fence riding. She rode Treasure, using Ron's saddle, which has a flank girth that the old ratbag didn't like at all. I rode Dawn, who was rather fresh, but as we understand each other better now, nothing frantic happened. Cora and another cow were extricated from the big weaner paddock and taken down to Moneymusk, a distance of 7 miles, and from there we rode round the Spring Paddock fence, fixing it up where necessary. We were also on the lookout for a few head of cattle, but as they didn't show themselves we came home, trotting and cantering most of the way, much to Dawn's delight; she loves coming home. On our return at 1.30 we found the kitchen crowded, with Billy Blakeney from Glen Eagle, Colin Lee Long, a Chinese carpenter with a splendid Australian drawl, plus his wife and father and two large dogs; there was quite a party going on. The Boss was hanging around looking thunderous; he hates having people drop in un-invited unless he knows them well, and Mrs Hassall was in a flap because it was Sharon's day off and I had been out riding all day and she had no one to help her produce food for everyone.

There was a heartbreaking accident at Conjiboy station the week before the Mareeba rodeo. Henry Wilson backed his truck out of the shed straight over his 2-year-old son, then came forward again without realising the toddler was even there. The funeral took place last week. It's a terribly easy thing to do, but poor bloke – he must feel utterly devastated.

I am making a dress out of two very bright tea-towels sewn together. The back has Waltzing Matilda on it and the front is the Australian Rough Riders' Association (ARRA) poster, covered with pictures of rodeo events. I shall add a tan pleated hem so that the end result is slightly more respectable. Trouble is, people will no doubt keep prodding me to stand up so that they can read the words of Waltzing Matilda.

There is a cattle sale in Mareeba on Thursday, so we'll be muster-ing on Tuesday, and I must go to the dentist in Atherton on the way. Fillings cost $5 each here, but the tax department will refund the cost at the end of the year. This seems to be a good way of working medical expenses; NHS take note!

7 August

As planned, we went mustering for the sale. Ron and I picked up a very suspicious mob that kept trying to break back, and as he was on Imp, the unreliable new pony who is apt to buck at any minute, Dawn and I had to deal with them, eventually yarding the stirred-up cows, where they chased everyone up the fence. They were drafted into seven groups with painted numbers on each lot: fats, stores, high-grade Brahmans, Droughtmasters, heifers, old breeders and bulls. Next day we set out in the truck with fourteen head for the sale, taking six hours to reach Mareeba as one or other of the beasts went down at intervals and had to be chivvied to their feet before they were trampled on by the others. At Atherton I stayed the night with the Wescotts, friends of the Hassalls, where there was a big party going on. Not knowing any of the people there, I took myself off to the pictures and saw a grotty film whose title escapes me. When I came back I had to camp down on the verandah, as all the beds were full of snoring guests.

The following morning I went shopping, read, drank milkshakes and after lunch trotted down to the dentist, who cursed and swore fluently but gave me a very efficient injection and three small fillings.

Ron has picked up a new saddle for Glen, with a flank girth like the one he has. She tried it out on Flax, one of her trusted horses, and was promptly dumped, saddle and all on the ground, much to everyone's surprise – including Flax's. The rest of that day was really rather wasted; Glen and I wandered around looking at things in various sheds, but neither of us did anything constructive.

Yesterday Ron and Glen took us jillaroos and the kids over to Glen Eagle to see the Blakeneys' new homestead. Only Billy and his sister Anne were there, plus a mob of workmen, and we had a noisy smoko in the shed in which they are still living, with a ton of lucerne at one end and the truck and tractor at the other. The new house is really lovely, built on slightly higher ground this time, and the new yards are completed to the satisfaction of all. After lunch we all piled into the Landrover and Billy took us for a scatty run along the river, where the men went pig shooting while we sat beside a dead crocodile on the bank. Not a pig was seen, so back we went. Colin Lee Long's large Great Dane and boxer-type dogs attached them-selves to me, using me as a leaning post whenever Billy tore down and up precipitous gullies at full speed. I felt decidedly dwarfed and very squashed between the two enormous panting dogs.

Today was the first really warm day since the onset of winter. I spent the late afternoon with pick and shovel, helping to dig a new drain for the septic tank – 6 feet down and 2 feet across. Hard yak-ker, as they say out here.

13 August

We went camping last week, Ron, Glen and I, in Wabble Creek Paddock. Sharon stayed at home to look after the kids and the Boss, as I won the toss to go camping, and before sunrise on Tuesday I had to get in the horses on Sunny, who was in one of her wild moods. When the horses started to run towards the yards, bucking and squealing in the early morning dew, Sunny of course made a feeble attempt to copy them and nearly fell over a small bush in so doing. I always feel most unsafe when I'm riding her; she has no brains at all.

I drafted off the required horses, had breakfast, rolled my swag and washed the separator, the horses were loaded by eight o'clock and we reached the camp site by ten. We set up the tents and had an early lunch before starting to muster by noon. Dawn jogged a lot – good for the tummy muscles after a meal – and as Ron left Glen and me to go a different way, we promptly took off our shirts and rode in our bras in the sun until about five. By this time we had come in a large circle and were nearly back at the camp, so we stopped our horses and plunged into a fairly clear waterhole. The water was waist-deep and refreshingly cool after all that sunbaking. Dressed again, we returned to camp only to discover the camp oven had been left behind (disaster!). The blade steaks and onions had to be fried on the shovel, which in fact made a most effective long-handled frying pan. Ron used his new portable transceiver to keep in touch with the homestead , and I had to say goodnight to Dave over and over again as the little boy was thrilled to bits to be allowed to speak on the machine.

I was asleep by eight in the warm night, requiring only a sheet for cover, then next morning we all woke long after sun-up and had a leisurely breakfast of cornflakes and bottled peaches, the remains of our supper steak, onions, tomatoes, eggs and sausages – not a bad effort, considering the only cooking utensil was a shovel. Later we trucked the horses to a dam and mustered the far side of the paddock, and just as we were leaving the dam a string of brumbies came in to water, being pushed by a beautiful shining black pony stallion. Ron shot one of the mares, alas, and the others galloped away, much to my relief.

Once again I rode shirtless, as I was sent up the creek alone; my back is shades darker already. Returning to the dam for lunch, we spread the tents out to dry off completely. A mob of curious weaners came along and chewed at the canvas and snuffled at the flames of the fire. Ron said, 'I wonder what they'd do if I got under the tent and said boo to them'. This he promptly did, crawling rapidly towards the cattle under the white canvas. They went tearing away, uttering startled grunts, and Dawn nearly took off too. She was *horrified*; the expression on her face was really something.

Later in the day we loaded the horses and went home, tired and tanned and content with our lot.

On Thursday Sharon and I went out with the Boss to pull down an old fence, roll the wire and pull the posts out to bring home, as fifty posts were needed to cover the new lav. trench Ron has been digging. Inside one post was a 6-inch orange and green centipede that we gingerly placed in a tin to take home and show everyone. Dave calls them 'canterpedes', a much more descriptive name than the real one.

21 August

I seem to have spent most of this week sitting in the sun painting yard gates with sump-oil; twenty-seven done to date, and there are around fifty in all.

One morning when only Sharon and I were in the house – the Hassalls had gone to Cairns the day before – we awoke to find Al was back from Mt Isa rodeo. I cooked a vast meal of steak, onions, eggs, rice and gravy for him, as he hadn't eaten for three days and was understandably irritable. Still, I did manage to get out of him that Doug had drawn and successfully ridden the feature horse to earn himself 50 dollars; also that Doug is not coming to Oak Park races, as Longreach rodeo is on the same weekend. So, that takes care of that. Oh well. *C'est la vie* I guess, and maybe it's for the best.

The Hassalls returned on Wednesday evening with a 9-year-old boy called Michael, who has been dying to visit a cattle station and reckoned he could ride 'very well'. Glen took him out on Sunny, but didn't let him leave the big yard as he really couldn't ride for toffee, but he enjoyed it all the same and still thinks the life is great. (I tend to agree.) These last few days he has been out on a resigned Sunny – on the leading rein – to help get the milkers. I had to ride her to bring in the horses one afternoon, and after shying violently at a chook that flapped out of the shed declaring it had laid four dozen eggs – these Aussies are all the same, full of tall stories – she took off at full gallop. Then, when we found the horses and circled round them to face home, she ducked her head and did three disconcerting kick-ups, snorting and groaning as if in great pain. After she'd been unsaddled she bucked and twisted all the way down to the trough, while Michael watched with open-mouthed horror.

I spent nearly all day Friday in the kitchen, and as it was such a lovely morning outside and I was itching to be in the sun, I became rather cranky, having to make a vast quantity of gingernuts, a Christmas pud for the races and some vile coconut biscuits that the Boss adores.

Sharon and Glen went out to get a cow with a new calf that hadn't suckled and was nearly dead, and they had to milk the huge fat Brahman in the crush with leg ropes at each corner, Ron holding her tail and her head in the bail. One tit was full of blood and obviously sore and the other three were too big for the little calf, being swollen with milk and hot with tension, so it was an ordeal for all concerned. However, they managed to milk her out eventually.

After Sunday lunch I went with Mrs Hassall to a christening at Minnamoolka – I really went along to help her change a tyre if she had a puncture. We got there about two. The day was clear and boiling hot and the first person we met was Al, with Harriet Scott, the Gunnawarra jillaroo, who is a super girl. Al was wearing a ridiculous little pork-pie hat he'd swapped with some bloke for his rodeo hat in a drunken haze at Mr Isa; no doubt he regrets the deal, for the new hat doesn't suit him at all. Brother Bill conducted the service in fine style, but the communion afterwards tended to go on and on, and as everyone was gathered in the drawing room with all doors and windows open to the verandah, the chorus of snores from the three station dogs sprawled outside added soporific interest to the proceedings. Tea was served at last, complete with luscious cakes and sandwiches, but Al, Harriet and I missed out on the meringues by being too polite; we should have known better!

Al wants me to go to the Ingham Bushman's Carnival with him on 10 September, but there won't be time, what with mustering, the races and Mrs Hassall's forthcoming trip by Redline coach to Darwin and Alice Springs. I have written to book a passage home on the *Oronsay*, to Australia House to return my passport, to godfather Billy Hawkes to see if I can stay with him in Ceylon for a fortnight and to Peter Rich in Sydney in the hope of staying with him before sailing. Glen wants me to see her married sister Rome in Goulbourn – she has a sheep property, and although I'm not in the least interested in sheep, I'd enjoy meeting Rome if she is anything like Glen.

29 August

Thank you for the postcard of Highland cattle. Everyone here was vastly amused by the shaggy old hat-rack of a cow and made rude remarks like 'How do they ever get fat enough to sell?' and 'The skin might just make a good blanket'. Cheeky beggars! My tea-towel dress *is* a mini, but not quite as bad as your drawing would have it. I wore it during the day at Oak Park after the races, and everyone was following me around trying to read the words of Waltzing Matilda – as if they don't know them already! So you have had a drought. Goodness me, how long did it last?

Well now – the races. The jillaroos sat in the back of the truck with the kids and Rum the dachs on top of all the gear, setting off at 8.30 on Wednesday and stopping for lunch under big shady pine trees at Bundock Creek. On arrival at Oak Park we landed on Spring Creek's camp, as we were to be their guests, and Sharon and I shared a big square tent with one other girl. After a supper of salt beef and boiled pumpkins we changed and went up to the 'dance hall', a corrugated-iron erection with no sides and a rough floor. It was freezing cold so we bundled off to our bunks at 10.30, but I couldn't sleep because my teeth were chattering so much. Finally I dragged the canvas swag cover off the ground and huddled under that, burrs and all.

We had steak for breakfast, then wandered around aimlessly all morning. In the afternoon we went with Ron and Glen to get a load of sand from the nearest creek to make a sandpit for the kids to play in at the camp. Supper was barbecued pork round the fire, then we went to the dance dressed in our 1920s outfits. Sharon won the prize for the most humorous character, dressed as 'My Country Cousin' in scuffed riding boots and a tatty skirt, with a splendid, stained satin ribbon round her hips. Loud applause greeted her as she received her unexpected award in confused amazement. I had a fair number of dances with nobody interesting, then seeing two boys we disliked descending upon us, Sharon and I tagged on to Ron and Glen, who were leaving, and escaped with them – much to their amusement and our relief.

Next day a nice nurse from the Northern Territory gave me my second tetanus jab – Glen made me bring the toxoid with me – then we went up to the first of the races, on which I lost 2 dollars and won 40 cents. Sharon and I became a bit bored with the action by lunchtime so we didn't go up to the course in the afternoon; instead we slept and got ready for the dance that night. Long after dark, when the dance was in full swing, Al turned up, surprising us all because he had said he would not be able to come. As it was he mustered all day and drove over after sundown – a good five-hour trip for him – and spent the rest of the evening dancing, mainly with me. Billy Blakeney, whom I rather like, also had one dance with me and asked for another, but Al had booked it firmly. Later he took

Sharon and me to a fireside party where Billy was playing the guitar, and we all sang and dozed in a cordial atmosphere in the warmth of the eucalyptus fire.

Next day after the races, however, Al 'got on the grog', as the saying goes, disappearing for the duration until the Race Ball, when he turned up looking decidedly the worse for wear. I had one dance with him but refused any more, as I was merely being used as a pit-prop to get him round the floor. He became cranky and strode out somewhat erratically to sober up on a bottle of Holbrook's sauce, returning to find Billy sitting harmlessly beside me. We had a blazing row a bit later, when Billy was acting MC, I telling Al what I thought of his behaviour, etc., and he saying 'Why?' monotonously at every other word. Neither of us got anywhere and in the end he sloped off saying he was going to fight Billy. I was horrified and was shaking all over when Ron and Glen came to my rescue and calmed me down. They told Billy about the proposed fight, but he only laughed. Al came staggering in an hour or so later – mud on his face where he'd fallen down – glared round the hall and disappeared again. Poor lad. I wish I could help him, but he just won't accept it from anyone.

The rest of the night was very enjoyable, if dampened somewhat by that incident, and at 2.30 Sharon and I went to relieve our feet by changing into riding boots and jeans – as did most people – returning to dance the night away until daylight. Then came the nasty job of packing up and saying goodbye to all our newly made friends. We left at one o'clock, Sharon and I sitting at the back in the blazing sun, where we dozed uncomfortably on and off while our hair and eyes steadily filled with red dust. When we arrived home, Al's car was at the gate; my heart sank as you can imagine. Not a word was spoken until after supper, when he asked to see me alone. He apologised very sincerely for being drunk and behaving in such a stupid and juvenile way at the dance hall, so now we are the best of mates again, thank the Lord.

On Monday we all walked around like a lot of zombies, automatically doing our chores, including an enormous wash that took three hours. Today I have been pruning peach trees and sweeping out the feed shed. Ron and Glen have gone down with some sort of bug and are laid up in bed for a while – probably a result of the rather stagnant river water at Oak Park.

5 September

We have been out camping for a couple of nights this week, which is all great fun; hard work and hot sun – my idea of bliss.

One evening after unsaddling Glen suggested we go for a swim in the dam, so we rode the horses bareback down to the water. Glen and Sharon went in first and swam their mounts across the dam, followed by Ron, while I plucked up courage to plunge Damona in. Finally we got soaked and all started laughing and Damona swam across and squelched out at the other side through the thick mud, with me clinging onto her slippery coat for dear life, weak with laughter.

During supper Ron told us that earlier in the day a goanna had come bustling through our tents and run up a tree, taking Sharon's face flannel and one of Ron's socks with it, neither of which were seen again. Glen then told us that at one of their camps a large bullock ate one of her bras, a *Reader's Digest* and a pair of scissors. I thought it was only sharks that ate such things.

After a freezing bath behind the truck, using a billy, I went to bed under *eight* blankets. The horses we needed next day were grazing round the camp, champing and thudding and whinnying softly all night, pleasant sounds to go to sleep with.

On Thursday, having snatched a quick lunch at the house, Sharon and I mustered some horses from Billygoat Paddock. We found them spread between the creeks, and started them homewards at a canter. All went well until two mongrel horses came galloping down the hill behind us trying to join the mob, whereupon Sunny attempted to look both ways at once, ending up doing a tight circle with her head under my chin, then galloped off at a tangent, out of control for a while until I steered her over to Sharon's horse, when she calmed down a bit. She'd never make a racehorse unless she wore blinkers. Let's face it, she'd never make a racehorse, period.

On Friday I did a huge wash, made ice-cream and a Christmas pudding, mowed the longer parts of the lawn, ate mounds of ripe mulberries and we had a birthday party for Ron in the evening.

12 September

Ron and Glen took us girls down to Cairns for the Amateur Races and Bushwhackers' Ball on Saturday. I milked at 5.50 a.m., raced the milk through the separator – almost producing instant butter – bolted breakfast and we set off in the VW at seven with the kids, picking Al up in Garnet, as the poor fellow had busted a pinion bearing. At least his car had! We had lunch on the Tableland, where the kids were left with friends, and got to Cairns in time to see four of the races. The grain-fed horses were terrific, so different from the grass-fed stockhorses out west at the picnic meetings. The liver-chestnut quarter-horse stallion whom Bits visited at Spring Creek ran twice and won; he's a noble beast and beautiful to watch in action. At the Cairns racecourse there is a special sandpit for the sweating horses to roll in after each race – an excellent idea, which they really seem to appreciate. We then went to a motel for a drinks party, raced along to the house of Don Lavers, the vet, to shower and change for the ball, and roared off downtown again to the Beachcomber for dinner.

It was really funny. The four of us sat at one table while Al went off elsewhere, and they took so long to serve us that Ron cut up the table decoration, which happened to be a proper pineapple, as opposed to a plastic one, and we ate that between us to quieten our rumbling stomachs. The people at the next-door table followed suit, but the waiters were well up to us, for on the bill at the end was added: 'One pineapple – 30c.'. One slightly merry jockey near us insisted on having his chook bones wrapped up for his dog; and a sulky waitress obliged with an expression of such distaste on her face that we were reduced to helpless giggles and silent tears. Her total lack of humour was just too much. Actually our meal when it arrived was quite something: avocado pears, chicken Tahiti and tropical fruit salad full of passionfruit pips – 'like the Meadowbank septic tank', as Ron observed, for passionfruit seeds always rise to the surface even after digestion.

On that note we left for the ball, which turned out to be a bit of a disappointment – to Sharon and me anyway. Most of the people there already had partners, and the men who were on their own only wanted to drink. Al danced with me once or twice but seemed preoccupied with troubles of his own, then Ray Cowan, whom we met at the Garnet races, came along like a breath of fresh air, but other than those two I only danced with a strange chap – a ringer from near Cooktown – and John Huey, who bought Sam, one of the Meadowbank bulls. John has a reputation as long as his six-foot frame and is not exactly my cup of tea. It's alleged that he once advertised for a governess who came out to his place to find that he had no children and no wife. We left about 3 a.m. and set off back up the range, reaching Sharon's place about an hour later. Ron picked us up next day and took us round Lake Barrine and Lake Eacham, two beautiful, flooded craters in deep rainforest, then we went home, dropping Al at Gunnawarra first and meeting Septic the grey stallion, whose real name of course is Septimus.

Yesterday I rounded up the horses on Sunny, who was even fresher than usual; she did a sudden prop and nearly disappeared from under me, then immediately reared up on her hind legs and I lost a stirrup trying to stay on board. Glen, who was passing, called out reassuringly: 'Don't worry Sue, she'll always come down again!'.

19 September

I doubt if I'll be doing anything much on my twenty-first, other than having a party with these pleasant people here: Glen, Ron, Sharon, Tammy and David and the Boss. Mrs Hassall won't be back from her bus trip by then, and Don and Doug will be miles away at some wretched rodeo. I might even go mustering all day, as I did last year. Glen tells me my riding has improved since coming here, and I must say I would rather be on a lively horse like Dawn, who races everywhere, than on a quiet, reliable nag like Damona, bless her shaggy old heels.

Last week the Boss hit his eye with a chip of basalt while clearing a new track, so retired to bed. Ron felt crook, so Glen, Sharon and I had a quiet supper all alone. Afterwards Glen developed and printed some of her old photographs, some of which she has given me for my album.

Immediately after lunch on Thursday we mustered the bulls. I was on Dawn, who was as fresh as ever and did little rocking bounds everywhere, looking at me out of the corner of her eye to see what effect it was having. The agent came with a buyer who wasn't sure if he would buy any bulls or not, but said he would return on Sunday. He decided to buy six, unbeknown to us, and the Boss was on tenterhooks the night before. He snapped at everything Glen or I said, vowed we wouldn't have time to get the bulls up in the morning and went on to such an extent that in the end poor Glen ran the horses in at ten that night by moonlight and drafted them alone, keeping the required ones in the yards all night. As it was we had yarded the bulls by nine and the buyer didn't come until eleven. I went down to the yards to release Dawn, just in time to see Sunflower, my favourite Brahman bull, chasing the long-suffering agent up the rails.

Yesterday we did a rush job in Wabble Creek, leaving at daybreak with the horses in the truck and getting home late the same day. We split up on arrival and drove any cattle we found in front of us, meeting up at the top and camp-drafting the thinner weaners out of the mob of 500. It was a really hot day, but all went well until the drafted mob started to move off down the spring. I tried to cut a corner to block them when down went Damona, up to her chest in bog. She struggled frantically, so I stepped off on firm ground and hauled her up, none the worse for her fright, and remounted. The cattle turned up the hill, decided they wanted to rejoin the others and broke in all directions. We hadn't a hope of retaining them all, though we managed to save about half the mob, and watched the others disappear – the usual trouble with trying to do things in too much of a hurry. On the homeward trip eight brumbies galloped down a gully off to the flank of the mob, and of course the wretched weaners happily followed them until they were forced back on course by cursing riders with stockwhips. After a bit more cutting out at the dam, Glen and I took them home while Ron and Sharon had their lunch at four o'clock and followed us in the truck. I cooked an enormous meal of steak, cauliflower, pumpkin, cabbage, spuds, onions, broccoli, white sauce and gravy, which we fell upon and devoured like starved dingoes.

25 September

I have been struggling with the new milker, Cora, who gives about two gallons of milk. Her udder is so full one needs two hands on each tit to make the milk flow well. For the past two days I have milked her and Wiggle, who has just been weaned and gives nearly a bucketful too, so I am certainly getting lots of practice at milking. Wiggle's only trouble is that she runs into the bail and gobbles her feed so quickly that I am in a lather trying to drag the milk out of her before she finishes her lucerne and hangs up on me.

Last Wednesday two simply dreadful salesmen rolled up, flogging a medical book and a set of Bible stories for children. Oh dear, how they did go on. Everyone had gathered in the kitchen where I was preparing the meal, and these men tried to drag me into the discussion too, but I stubbornly continued cutting up onions on the next table and pretended to be just part of the works. Ron was in fits, controlled with obvious difficulty, the Boss walked out scratching his head and Glen listened with a very solemn face, which made us laugh even more. Sharon was running the horses up, and when the two salesmen saw her cantering past on her own horse, Tiffany, they said 'Can she *ride*?' in awe-struck tones, as if they'd never seen such a thing in their lives before. It was all of two hours before they finally departed, leaving us to explode into bottled-up laughter. Glen deserved an Oscar for her performance.

It was Sharon's twentieth birthday on Thursday, and we had a little party. I gave her a pendant of a local pink pebble on a chain, which I think pleased her. On Friday I oiled more gates and watered the young trees in the yards. In the afternoon I fetched the horses when they came to water at the troughs. I tried using Ron's stockwhip and succeeded in tripping myself up twice and nearly taking off my left ear. They're devilish weapons in inexperienced hands.

On Sunday morning a cow was shot, and we butchered her on

Monday after breakfast. I de-boned the neck, removed the kidneys, skinned the feet and wrapped and labelled all the cuts, then spent an amusing time once again making sausages. I helped the bangers out of the mincer and learned to twist them fairly efficiently so that they formed clumps of four (more or less) in a long string. Later I rode Sunny bareback to fetch the cows, ripping my shorts while mounting, and found Wiggle well ensconced in the bog and the wild brindle bush-cow running in wide circles with her head up, looking very fearsome; so the usually peaceful chore of bringing in the milkers turned into a minor sort of rodeo that night. I went on Sunny because frankly I'm a bit wary of approaching that old Brahman on foot. Her torn udder is getting better slowly, and it will be a relief when she goes bush again.

3 October

Here it is, October already, and only a month or so to go before I leave Meadowbank, when emotions will be torn to shreds once again. I'll miss these good folk, a lot of the horses – even bitchy old Bits – and the wild Brahman cattle, but it's good to know that I shall be going to the Swinbournes' for a holiday and not to total strangers.

On Wednesday Sharon and I took a mob of weaners out of the yards down to Bower Bird Mill. I rode Dawn and Sharon was on another lively horse, so our return trip included lots of trotting and cantering to keep the steeds happy. As it was a fine sunny day we rode shirt-less to further our suntanning programme. Glen and I have started to milk at 5.40, as daylight comes earlier these days, and I have finished the fifty-first gate in the yards, covering it lavishly with sump-oil, so now they are all gleaming black and slightly more weatherproof than before. After that, wire brush in hand, I went up onto the shed roof and brushed the old paint off the lettering M-BANK (there for the guidance of aircraft) and am going to repaint it some time this week. It's pretty hot underfoot up there, but the wind keeps one cool enough. The warm weather is on the way at last, though I still have three blankets and two heavy cats on the bed. The grass has dried up rapidly, but the cattle are in excellent con-dition, apart from a few weaners that are being supplementary-fed. The horses are getting hungrier, too, though still fat and sleek, and reach over the fence to eat the mulberry leaves and Glen's dahlias.

Don and Doug are at Wendouree doing some fencing in Greentrees Paddock, where the thirty-odd head of brumbies run. Don is training a horse to be a roping pony for Doug to take round with him next year instead of Twinkle; Don has retired and thus has more time on his hands. He can't settle in one spot, though; he's off to Victoria later to pick fruit.

10 October

Sunday was a pleasantly unusual day. Al brought Sam, the head stockman from Gunnawarra, and Harriet Scott, their bookkeeper-cum-jillaroo, to spend the day here; first the Boss took us all in the 'rover to see the latest crop of calves, the weaners and preggy cows, and then we returned for lunch. Sharon then ran in the horses and the five of us – Al, Sam, Harriet, Sharon and myself – set out on horseback for Mt Lang, the highest hill around here for miles, from the top of which one is supposed to be able to see Townsville's radio tower. I rode Dawn up the steep and narrow path, and if she'd taken it into her head to shy we'd both have gone over the edge. However, she was fortunately on her best behaviour. There was a fabulous view from the top of the old volcano; you really can see for miles around. We cantered all the way home in time to meet Ron and Glen, who had returned from a trip to Ravenshoe, and they took us and the kids down to Scrub Dam in the 'rover for a swim and a barbecue just before sundown. The dam is large and very deep. I swam a long way out to a tree in the middle, where the water was dangerously cold under the top eighteen inches or so. I'm not surprised that so many people are drowned in dams, the victims of cramp with the sudden cold water. The barbecue was steak, of course, charcoal-coated spuds with masses of butter, and raw carrots.

The other day I did a bit of bareback work on Sunny the nightmare, putting the wild brindle cow out and bringing in *Elbat*, the new milker heifer from the *Table*land. It's like sitting on a Volkswagen beetle on board Sunny. There's nothing in front of you, as her shoulders slope away dramatically, and if she stops suddenly, which is not unusual, one tends to find oneself either up on her ears or sliding along her neck to the ground.

Glen gave me my third tetanus injection today, chasing me round the kitchen table with syringe poised until she cornered me by the woodbox. After the dire deed was done she was shaking more than I was as she has only ever done cows before.

16 October

Thank you for your letter, received on Saturday when I 'went to town'. Fancy having snow already, even if it is only on the high hills. We had our first storm of summer yesterday, which let down 56 points of rain and some hail. It came at just the right time, as the grass was almost dead, and now everyone is expecting early storms. Sometimes it doesn't rain until after Christmas, by which time the cattle are falling away and thousands of dollars are spent in keeping breeders and weaners alive with imported feed and carted water.

I had a highly unusual twenty-first birthday. I slept in until 6.30, awaking to find a parcel of *raw liver* on the pillow. Glen had asked me what I wanted for my party, as the roosters were too skinny to kill, and although I would never reach the stage of being tired of steak I suggested liver for a change – and got it! So breakfast and lunch consisted of just that, fried in butter both times. I did some washing later, watered the trees in the yards and had the afternoon off, while Glen iced a cake for supper and set up the table on the verandah with an arrangement of ginger flowers and sprays of white jasmine on a deep red cloth with white Lippizaners on it. A road of blue toilet paper led to the cake, on top of which was a signpost saying 'Anakie' and a cardboard tent with one candle outside it representing the campfire. A hole had been cut in a corner of the cake and a windlass of wire set over it, with a wire figure of Sharon winding it and me further away hard at it with a pick. On the windlass was a rope, at the end of which was a gold charm. Although we're hoping to find sapphires on our prospecting trip, not gold, I thought it was a brainwave on Glen's part. The cake was coffee sponge – my favourite – with boiled coffee icing and chocolate writing on it. They drank my health and sang, then we had a lovely meal of steak and pineapple casserole, cauliflower cheese, carrots in butter, beans and chokos, ice-cream and apple pudding. The Boss took his into the kitchen to listen to the wireless, so we 'young' had a nice noisy party to ourselves. Sharon has given me a super hand-tooled leather belt, the kids gave me a couple of Aussie-designed tea-towels to make a shirt or anything else I fancy, and Ron and Glen say they have been trying to get a cut local agate of some kind for some time, but without success. There is a little Rocke Shoppe near Garnet, and they're going to try there next. It was a lovely party, and I was deeply touched by all the thought that went into making it a memorable day for me.

The following day was definitely an anti-climax, improving only slightly when Sharon and I went boundary riding for four hours in the hot sun. As it was Friday the 13th I had visions of trees falling on us and snakes attacking from all sides, but nothing untoward happened, and the fence was only broken in a couple of places. We rode without shirts as soon as we were out of sight of the homestead and are nicely brown again. After lunch we mustered the horses and cantered them all the way to the yards to get two – Dewdrop the black mare and Dandy, a red bay gelding – for Harriet, who has come over from Gunnawarra to help muster Beasleys Paddock this week. Unfortunately it's my turn to stay behind and look after the kids and the Boss, and Harriet has come to take my place on the camp. They left yesterday and got soaked in the storm, so they said over the portable transceiver last night. Ron has a fishing line and shanghai with which to put the aerial up a tree for better reception.

Tammy has boils at present, poor wee girl, and becomes understandably cranky at times; Dave is growing up fast and is a super little boy. Today they were up in their 'Lucerne House' (the hay shed) after a smacking fight that lasted only a couple of minutes without tears. Thank goodness they both have ravenous appetites and are no trouble to feed. Presumably all this is good practice for me for the years to come.

Saturday the 14th was a simply splendid day – I haven't enjoyed myself so much for ages. Ron had to go into a Lodge meeting in Garnet, so he took Sharon and me into town with him, where we met up with Harriet and I had my first legal drink in a pub (a shandy). Then we three girls 'did' the town, spending lots of money in the jukebox and on milkshakes at the cafe before moving on to the cinema to see a ridiculously funny film called *The Incredible Mr Limpet*, about a man who changed into a fish and helped the US Navy detect submarines. It had us rolling in the aisles for some obscure reason; perhaps it was just the silly mood we were in that made it seem such an entertaining film. The picture house is simply a corrugated-iron shed, and you have to be a regular to know where not to sit, as when it rains the roof leaks like a sieve. The seats are the same as at Wandoan, canvas double deckchairs in which one reclines and is just able to see the bottom of the screen, while one's legs get pins and needles as all circulation is rapidly cut off. We emerged just before midnight, sat in Harriet's car for another hour or so waiting for Lodge to finish, then came home and got to bed about 3 a.m., after a shower. At 5.30 Sharon's dreadful alarm clock jangled us awake – and so the Sabbath began.

23 October

On Friday, you'll be pleased to hear, Damona jiggered my left knee again, and it is still twice the size of the other one, with a handsome black and green bruise. Today is the first time it has really taken any weight, though I still collapse periodically, like Sunny with a weak tendon. We were cantering towards a dam to move cattle away, and I gave Damona plenty of time to realise we were going to the left of the tree in front. However, she went to the right, ramming my knee into a pretty sturdy ironbark. The impact unseated me and I clung round her neck until she stopped rooting around in fright, then got off to see if my legs still worked – which they did. I therefore continued riding for the rest of the day, taking the spare horses 6 miles home and returning to help with the mob of 628 weaners and young steers. However, when we eventually got back and unsaddled I couldn't walk at all, and have been laid up ever since. A stye and a gumboil are also beginning to make their presence felt; it never rains but it pours, or something.

On Tuesday morning I had the house to myself, as the others were still camping, and the Boss took the kids off my hands. After lunch I rode out to help the others camp-draft the steers, riding Echo, the Boss's old mare, whom nobody likes riding as she invariably comes home lame. Later I went with Glen to bring the horses home, and Echo jolly well rooted with me in the creek; however, it was easy enough to stick, as she's really no more dangerous than an armchair with hiccups. Sunny was in the mob too, and she was so funny to watch in among the others, squealing and reversing rapidly to kick, then floating through the air with her nose way up until she tripped over a pebble – behaving like a mentally defective stallion, in fact.

Next day it was my turn to go camping and Sharon's to baby-sit, so Harriet, Glen and I took the horses down to Wabble Creek. Harriet taught me all about trotting on the left and right diagonals, which I found quite amazing; it never entered my head that every horse has two ways of trotting. It was a hot, windless day and there was not a drop of water in the holding paddock. As we mustered on up the spring we found one dead steer, a dead weaner and the skeleton of a 'roo, so things must have been pretty desperate recently. We were back at camp by four o'clock to meet Ron with the gear, then I went off alone and picked up twenty-seven head before dark, when there was an eclipse of the moon.

Next morning Harriet found a 6-inch 'canterpede' under her shirt, which she luckily spied before donning the garment.

Harriet and I slept that night with the sides of the tent hitched up, as the moon was so lovely, and got up at four in the morning, like all good drovers, getting the cattle on the homeward trail by a quarter past six. I remained behind, took the old Morris up to Brumby Paddock, spread the tents out to dry and waited until Glen came with the horses, when I caught Damona and went down to the dam to move the cattle. That was when my knee copped it.

It is now after lunch and a storm is about to break; the old windmill is clanking round at a rate of knots and the birds are huddling under leaves and keeping their beaks shut. It looks as if it might hail; the sky is very green and ominous.

Last day of October

The knee has just about recovered. I rode most of yesterday, and it was only a little bit rickety on dismounting.

The Boss has arrived back with Mrs Hassall, who has just finished her six-week bus tour. She was most impressed with the Northern Territory and Ayers Rock, though she says three buses and one aeroplane arrive *every day* at the Rock – what a ghastly thought! She showed her movies of the trip after supper. They were very good, but I don't think I'd like to travel that distance in a bus.

We all went over to Whitewater for the day on Sunday, and I went on with the Boss, Ron and Glen to the graziers' meeting at Mt Surprise station, which went on from ten until four, with lunch served by the horsebreaker's wife at noon. The meeting was interesting, all about dip-resistant ticks and aerial dog-baiting, but it tended to go on a long time. The room was very hot, and my feet started to swell after sitting for hours. That evening Ron, Glen and I went to Boomerang (our western neighbour) for supper, and sat from seven-thirty until ten looking at slides of Mrs Fanny Adams in her white hat attending various weddings. Oh dear, we were *so* bored. If it had been three hours of cattle and horses it might have been different, but weddings! However, we had a chilly but amusing journey back in the 'rover, reliving the slide show with many alterations and amendments.

Yesterday both of Dawn's shoes were loose – only the front feet are shod here – so I had to ride dopey Damona to cut the bulls out of the breeders. Glen and I met two running bulls, with Ron galloping at right angles to block them, so we joined in to help. Kismet, a seven-eighths-bred Brahman, turned nasty and came for Damona, who needs a rope and pulley to turn her round, and Ron raced in with his whip shouting 'Look out, he's fair dinkum!'. Kismet evidently has a reputation to live up to, for later on he came out of the mob and attacked us again, but this time Damona actually saw him coming and trotted up to the other side of Borax, the boss bull of the herd, thus thwarting Kismet with her intelligence – which is far superior to her manoeuverability. Half an hour later I nearly hanged myself in the scrub. I was pushing a small tree out of the way when down fell a thick vine and Damona walked on until I was almost dragged off by the neck before she realised something was wrong and stopped. I have a fine red graze on my neck today, a good topic of conversation for suppertime.

5 November

I wrote to you last just before going to muster fifty cows and calves with Glen and Sharon. We were riding along shirtless, half asleep in the hot afternoon sun, when Dawn trod on a stick, the end of which sprang up and struck her in the belly. Boy, did she take off – at 170 miles an hour, Glen said later – leaving me behind in a surprised heap on the ground, having first been dragged a couple of yards which relieved me of some four square inches of skin from my back. She bolted for easily half a mile before she realised that the branch wasn't actually chasing her. Glen retrieved her, having told me to stay where I was as I looked so comfortable reclining on the dry ground.

On Wednesday we three girls went out again, checking fences and looking for seven head of missing cattle. I'm afraid I lost my hat – or rather, your planter's hat, Dad. It was attached to the saddlebag, because I rarely wear it these days, and when I looked it was gone. I *am* sorry.

On Friday we mustered Junction Paddock for the young bulls and were each given a box of matches. Every time a match was struck Dawn leapt forward at least three feet – she's such a nervy beast – but I started a fair number of fires in the dry grass as well as wasting many matches, and soon the bush behind us was burning merrily, with clouds of smoke billowing into the blue sky. These fires are necessary in this type of country to clean up the undergrowth without damaging the trees much, and to start the seeds of the native grasses germinating after the next rain. If the small stuff isn't burnt every year it soon builds up, and then there is the real risk of uncontrollable bushfire, which can be disastrous.

The Honnerys from Boomerang came over on Saturday for the day, and in the afternoon we took them down to Scrub Dam for a gorgeous swim and an early barbecue, cooked mostly by me. As the tea billy had been forgotten the tea was made in the camp oven – stewed tea? We went home at 7.30, as the Honnerys wanted to show us more slides – oh dear. We sat through a further two hours of weddings and underexposed scenery. Ron, Sharon and I were nearly asleep on the woodbox with the cats, but Glen kept up a splendid act of being interested, though even she began to flag towards ten o'clock. But the Honnerys are nice people for all that.

14 November

My old ailments have all cleared up, thank goodness. Things have to look after themselves out here – if you can walk you're fit – but now I have a new and painful boil developing on the site of an old mozzie bite on my jawbone, making all my neck glands sore. It's a damned nuisance because I'm all packed up and ready to leave, depending on the weather. Anyway, there's a cyclone rapidly approaching Cairns, which gave us a bit of rain here, so it may be another week before we can get through to the bitumen road. I have had my last ride, alas, and washed the girth and saddle blanket today. Things are definitely winding down.

Last Tuesday we mustered Scrub Paddock and everyone was terribly cranky for some reason, riding most of the day in silence, which was most unusual and thus a bit depressing. The flies and any beast that happened to stick its nose out of the mob were cursed roundly, but otherwise no human voice was heard. I tried to hold a mob alone, but they just wouldn't stand and spread everywhere until I nearly lost the lot to the scrub. Happily help came in time. It was a hot, sultry kind of day, not at all pleasant, and black dust rose in clouds off the burned grass. Dawn jiggled all the way home until I smacked her soundly on her foaming neck. She nearly jumped out of her skin at the noise, but at least she stopped her ridiculous jogging and walked sedately the rest of the way.

On Thursday Sharon and I mustered Billygoat Paddock to bring in the three old, retired horses, one of which Tam is going to ride while her pony is visiting the Gunnawarra stallion, Septimus. It took us all morning to find the old wasters, then they were *so* slow going home, falling over stones and looking decidedly spavined, not unlike three moulting bathmats with patches of tangled hair hanging down.

The last two days I have been painting the sitting room pale green and typing for the Boss, while the rain poured down outside. Yesterday we brought in a killer, which will be shot tonight and butchered on Wednesday; then *if* it doesn't rain any more, Glen has to take the kids to the doc. for injections, so I'll probably go to Atherton with her on Thursday. The mighty 'Sunlander' will take me from Cairns south to Maryborough two and a half days later.

Christmas at 314 Queen Street

28 November

Here I am with the Swinbournes again. It's lovely to see them after all this time, but oh dear, your daughter has become very anti-social since emerging from the bush – it is really quite alarming. Parties and dances mean very little, and all the city folk one meets seem to be so artificial, always poking their noses into other people's affairs and rushing around like frenzied ants, getting nowhere fast.

I spent a miserable, lonely night in a guesthouse in Cairns. Meals were not provided, so I wandered up to the Chinese restaurant after Glen and Ron left me, ate a self-conscious meal at a table for one, then tried to sleep on a hot bed in a stuffy little room – and failed dismally. Sunrise was a great relief. If I had but known, Don Hafemeister was in Cairns that day too, on his own.

Now I am back in civilisation and finding it difficult to adapt. Jean took me to the Golf Club tea party where I was introduced to all and sundry as 'Miss Hawkes of Scotland'. Time dragged until a youngish woman came along and sat at our table. She said she had been droving up in the Gulf country when she was younger, so from then on our heads were together and we talked non-stop about our favourite subjects – cattle and horses.

Ruth Andersen, the girl I vaguely met last year at the dancing club, came round to see me one evening, which was nice of her, and we arranged to go out to a dance on Saturday in a little place called Tiaro. On Thursday Jean left to go down to Tweed Heads on the border for a few days, where Rod Swinbourne has the boat moored. He's living on it while working near by, and next year they plan to buy a house down around the border region. Terry and I are going to stay on the boat next week. On Friday I caught the bus into town and did a bit of shopping without getting bushed once; I can find my way from a fence to a certain creek over various basalt ridges, using the sun as a guide, but let me loose in a fair-sized town where all the blocks are square and the roads parallel and I'm totally lost, or bolt in the wrong direction.

After an early tea on Saturday, Ruth came to collect me at 7.30 in her little old Morris 1000 and we drove the 18 miles out to Tiaro. My heart sank when I saw the few dreary-looking males with glasses and spindly white arms, but Ruth introduced me to mobs of people, so I had lots of dances and quite enjoyed the evening after all. On Sunday morning the budgie woke me up demanding to have his cover taken off as he was missing the best part of the day. He obviously thought humans were the end, being so darned lazy in the mornings, and had no option but to leap up and down on his perch making as much noise as possible. Ruth came round after lunch to take me down to Hervey Bay for a swim. There were masses of people swarming everywhere, most of the blokes in souped-up cars screaming along the road and showing off their torsos to anyone who was interested (I wasn't). The water was murky and there were too many noisy folk around for much enjoyment, so we returned to a friend's house where a few people were turning up for a barbecue on the lawn. The father, a German, composes music and makes organs and pianolas in his spare time: a strange old man who gave us a 'musical evening', playing all his various instruments and singing. It was pleasant enough for half an hour or so, but he *did* go on. Most of us were glassy-eyed by the time he'd finished.

Yesterday Jean came back, exhausted after battling her way through Brisbane in the car, having done a trip of some 200 miles. She found me dyeing my bathing suit black, as it was looking somewhat grey from swimming in muddy dams and creeks out west. It's a dark blue colour now, not black, but anything is an improvement.

PS Letter from Glen. She found your hat, Dad, way out in Bowerbird where it dropped off my saddlebag, but it has been ravaged by termites so now has a lacey frill at the front. Peter Darvall came out to geld Araldite (a good sticker) and Daybreak, Dawn's colt from two years ago. Daybreak didn't recover from the knockout drug so is now as good as dead, completely listless and useless. Apparently this new dope has that effect on one in two hundred colts. The Meadowbank folk are naturally very upset, and so was poor Peter.

They have sold the big Brahman bull Kismet, not before time, and Ron cut his shin badly and had to have masses of stitches put in it.

12 *December*

The ABC has been telling of big snowfalls in Aberdeen: one foot in three hours, which sounds more like Canada than Scotland. Are you snowed in again?

It was a cold and windy week down at Tweed Heads, confining Terry and me to the boat for hours on end until we became a bit tired of each other's company. I took myself out to the Coolangatta Zoo one day, while Terry braved the surf; he doesn't like going around with me because I look like his elder sister, and as he is only 14 I see his point. The zoo is supposed to be the third largest in Australia, which doesn't say much for the others. It took me five minutes to walk round the cages and see a pair of lions, a leopard, a bear, a few monkeys, pigeons, 'roos and emus, a couple of Shetland ponies, a camel and one giraffe, not forgetting a cage full of black cockatoos looking thoroughly cheesed off.

When Rod Swinbourne came back from work we three went up to the windblown Golf Club for a shower, and later to dinner at Surfers Paradise at the El Rancho Bar-B-Q. The name was enough to put me off for a start, but inside we had huge T-bone steaks and a bottle of sparkling burgundy under dimmed lights with a band playing softly in one corner and a few folk dancing – very pleasant. I was decidedly dizzy when we left just before midnight to return to the *Sue* in the Mini Moke.

Thursday was spent on board scraping paint off the hatches and helping Terry put fresh paint on the rubbed-down surfaces. Later I bullied him to come to the pictures after a Chinese meal, he complaining all the way about only being 14 and having to go out with me all the time. The film was *The Great Escape*, which I have seen at least three times before, but as it is a good story and has Steve McQueen in it, it was no great hardship to see again.

Yesterday Terry and I took our leave of his father and caught the Pioneer bus back to Maryborough, changing at Brisbane onto the Cairns Express, which gave me a great longing to 'shoot through' to that distant city. However, Ian comes on leave on Thursday, Murray (20) will be up from Sydney soon, and the sun is warm here. You have no idea how awful the weather was down south; my blood must be as thin as metho by now, for I feel the cold terribly.

Merv Carruthers wrote and offered me his new 6-cylinder

Landrover for the prospecting trip, though I'm sure he will need it himself. He says he will fit us out with blankets, tarpaulins, etc., which is very kind of him, and has asked Sharon and me to stay with them at Wendouree for a while prior to going to Anakie.

19 December

I am beginning to cheer up a bit and enjoy life once more, though I still long for the bush. However, now that Ian Swinbourne is up on leave we are having quite a good time. He arrived last Thursday, much taller and more handsome than last year and full of his new hobby – parachute jumping. We went for a swim in the Mary River at high tide, when the water is about twenty feet deep, then ate iced watermelon on the lawn while drying off. After lunch Jean, Terry, Ian and I went down to Hervey Bay for another swim, after which Ruth came round to take me to see *The Incredible Journey* at the flea-pit: a lovely story about two dogs and a Siamese cat walking home across America. Returning at eleven we found Ian still up, so spent the next two hours looking at photos and talking non-stop.

Next day Jean and I went to a Christmas drinks party in a butcher's shop, no less! Standing among the chopping blocks and mincers were two 9-gallon kegs of Goldtop beer, being rapidly soaked up by the big mob of people all crammed into the shop. I drove the Holden home, having drunk less than anyone else, and after struggling with a belligerent mutton chop we slept until four, when Jean went to play golf with a splendid hangover.

When Ian returned from spending the weekend at the Bay with some friends, he took me to play squash with a husky, nice-looking midshipman of about twenty. They each played a couple of games with me, being most considerate of my newness to the sport, which is a fine form of exercise in that humid heat – better than a sauna bath any day. Ian and I then came back for a dip in the river, but as the tide was coming in fast and the current was swirling strongly I just hung onto the jetty and floated close in to the bank. Murray comes up this weekend, the son I haven't met yet, and Jean hopes we hit it off together as she wants to keep me in the family. I seem to be the victim of matchmaking out here.

The Perretts of Kabunga West, Wandoan, have invited me to go out there for Rodney's twenty-first on 6 January. If it is possible for me to go from Wandoan to Anakie without having to return to Brisbane then I shall definitely go, for it would be good to see them again.

Boxing Day

Murray arrived on Christmas Eve at about 4.30 after twenty-six hours in a bus and was naturally a bit tired. He is nice, though skinny, and looks as if he could do with a few weeks on corn to fill him out a bit. Jean gave a party here on Christmas Eve. Ruth came with all her dance records and stereogram and everyone danced, then we had presents off the tiny little artificial tree and sang carols and folk songs. At 1.30 the guests left, at which point Murray – a real night-bird – came to life and we danced until dawn. Rod was up for the weekend from the coast. He and Jean went to bed at 3 a.m., Terry and Ian flaked out about the same time, but Murray and I kept going until the sky was light and the birds were singing to herald a hot Christmas morning. Ron and Glen have sent me a lovely book called *Australia*, which is made up of photographs of all the different states, a fine present and quite unexpected.

Last week Ruth took me to the Bay with a few others and we wore *shorts* to a dance. I was horrified, wanting to wear a dress, but as everyone else was similarly attired, some even with bare feet, I did the same, though it did feel peculiar. The dance was quite enjoyable and the two new boys I danced with were ringers, one from Eidsvold and the other from Biloela, so we talked cattle all round the floor. It seems I have only one topic of conversation these days!

2 January 1968

I spent a night with Ruth out at her parents' place this week, a small dairy farm with a few bush paddocks and four tiny little dams. We walked round the whole place in two hours, meeting their one horse, which hasn't been ridden for three years (I wasn't going to be the first to try, either), two nice sheepdogs, a cat, a parrot, and a baby swamp wallaby who is quite lovely, with tiny, strong, black hands and huge ears. He hung on to my finger and accepted peanuts greedily. Mr Andersen had to wear an apron all the time the joey was small. He really felt a fool, but the little fellow firmly used the large pocket as his bolt-hole in times of danger. Ruth's mother is a sweet woman who works hard on the farm, growing vegetables and a few rows of groundnuts.

Murray and I did not hit it off too well, much to Jean's disappointment; I get on much better with Ian, for all he's a couple of years younger. On Thursday Jean sent Murray and me out for a drive, to our mutual horror, but in fact it turned out to be quite fun and we managed to talk fairly freely over a range of subjects and explored various dirt roads, where we found some pretty little fishing villages peacefully standing on stilts on different branches of creeks entering the sea. After tea the train left for Sydney, taking Murray with it. When we said goodbye he asked if I wanted to see some nightclubs when I'm in Sydney next month, so we're still on speaking terms.

That night Rodney rang through from Wandoan insisting I come out to his party on the 6th, saying that I could get a lift with two other boys who are going from Maryborough, and that he ought to be able to get me up north, somehow, in time to meet Sharon. So I'm going.

Last Saturday Ruth and I drove down to Hervey Bay for another dance. The first jazz waltz, when they turn down the lights, I had a vile, drunken, redheaded linesman and just about dislocated my shoulders trying to hold him off. Then a tall, fair, deaf cane-grower from Bundaberg with a high, inarticulate voice took me up for a gypsy tap; a small orange-grower's son had me up for a barn dance and a huge, tall Texan in a black ringer's hat gave me an Oxford waltz. The latter was in fact Australian, like the rest of them, but with his hat, red checked shirt and cowboy boots one could well be forgiven for taking him to be a Yank.

Sunday being Hogmanay, Jean and I had to go to a drinks party, while Ian and Terry took the car to another gathering. Our party was so stuffy and boring that when Ruth rang I gladly escaped, and we went once more to the Bay for the midnight to dawn dance, where the air was stifling with the huge crowd within. The tall Texan grabbed me for the dance right on midnight, when everyone sang and stamped and yelled until the whole building nearly collapsed, sweat dripping off every brow. We had a breather on the balcony with sundry other folk and watched the cars driving slowly up and down the street below, blowing their horns like mad. Among them was the Swinbourne Holden containing Ian and Terry, each with a small girlfriend tucked under his arm. Ruth and I went out to the cafe for a drink of orange squash and were pounced upon and kissed by at least twelve different boys, all wishing us a happy New Year. Then a crowd of us walked for miles along the beach, singing and shouting, and after a drink from a stray hose-pipe (everything else was shut), Ruth and I left as the sky was paling, getting home at 5.30, when we had breakfast and went to bed. I slept until 12.30, waking to intense heat, had a refreshing swim in the river, roast chook for lunch and another snooze for a couple of hours until it was cool enough to be sociable again.

So all in all 1967 was a great year, for me anyway.

Rosevale, via Calliope, Gladstone

11 January

Last Wednesday I spent the night with Ruth Andersen at their farm. I helped her father kill a vealer, skinned it and got the guts out intact, and also milked an ancient yellow cow called Heather, who stood quietly in the yard, no bail or crush or even leg-rope needed. The following morning was spent fooling about in the nearest dam with the two dogs and a canoe that sank every three minutes to the muddy bottom, where eels wriggled away from our feet. That night I returned to 314 Queen Street and took Jean out to dinner as a small 'thank-you' for all she has done for me; we shared a bottle of pink Barossa Pearl and became pleasantly giggly over the roast chook. In the morning Ruth came round and took me down to the Bay, where we ran into the tall Texan and his mates, still in their black stetsons, and all went for a swim in the clear green sea. One of the boys, Toby, was particularly nice, not unlike Don to look at, and we had a pleasant yarn about horses, as he has one of his own.

Friday was my last day in Maryborough and once again I bade a sad farewell to all my good friends in and around that pleasant town. On Saturday morning Paul Everett and his brother Stephen, who plays the bagpipes, arrived at 6 a.m. to pick me up for the trip out to Wandoan. It rained most of the way – seemingly the wet has set in early this year – so Sharon and I might have trouble on the black soil around Anakie later on. I drove the Everetts' Holden along the 50 miles of bitumen and the boys drove the remaining 250 miles on the dirt roads. They didn't like the latter much, being real city slickers, and were worried about the suspension on the corrugations.

Rodney was surprised and genuinely pleased to see me when we eventually rolled up at Kabunga West, where a shooting competition was in progress among the mobs of young folk wandering about. I noticed a bloke called Sam eyeing me over, and sure enough he bogged me down for the rest of the evening. It rather spoilt things; he was nice enough, but I'd have preferred to remain unattached. There was a massive barbecue after sundown, with dancing out of doors until it began to rain, when we all moved inside. At about three the party moved across the mud to the 'High Hut', where the revelry continued until daylight. I crashed at four and slept on the floor of the spare room until six, when I was awakened by the sound of reveille being played on the pipes. So were a lot of others, judging by the number of groans and muttered oaths. If anyone had had the energy they'd have dunked the piper, Stephen Everett, in the nearest water trough, pipes and all.

Some time the previous evening while jiving with Rodney I put my left kneecap out for the third time, and today it is still a little swollen and painful. At 7.30 I helped milk a very cranky cow, then we all piled into and onto the two A-model Fords – about ten people to each car – and careered off through the wet soil to swim in a dam, with mud splashing up as the cars did broadsides over the washouts. Why they didn't roll I'll never know, especially when they were so top-heavy, with folk hanging on precariously on all sides. Both radiators were boiling on arrival at the dam, where we had a good swim before returning to the homestead. One car ran out of water completely on the way back. Rodney scooped up some filthy, soupy mud from a puddle, poured it into the radiator and away we went, none the worse.

After breakfast at 10.30 we all went off to see Dick Perrett's new dam: a truly enormous hole gouged out of the red soil, as yet empty. Both Holdens became bogged at more or less the same place so we walked about a mile through mud and burrs to Bottle Tree Hills, only to find Dick away.

On Monday I was asked to help muster sale cattle, to which I gladly agreed, and was given a big, bay, camp horse to ride. Mr Perrett, Bruce (18), Ursula the nice daughter and I set off, and Mr Perrett sent me ahead to put some bulls through a gate. Only then did I realise what the big bay horse could do. He was wonderful, very fast and so keen on his work, spinning on one leg like a polo pony, his ears pricked all the time as he watched each beast, his big hoofs pounding about on the damp ground. We had lunch at a dam in the drizzle, then started camp-drafting in earnest. The cry went up,

'Sue, get that big roan bullock!', so into the mob I went. I pushed the beast to the outside of the herd, then chased him away, keeping right on his tail where possible – and believe me he didn't go straight for a moment, as he tried to double back to his mates. Later a heifer broke and the bay took off after her at a flat, powerful gallop. She propped, Silverstream spun round in the mud and we had her mastered. That is the first real camp-drafting I have done, and on a good horse it's terrific. It was the fastest work I'd done for a long time, and I was pretty stiff that night. As we took the sale cattle home it began to rain, but Rodney drove out to meet us with coats for all.

A very nice 19-year-old called Bill Stevenson was staying at Kabunga West with Rodney, and he offered to take me up to Rocky in time to catch the Midlander, as he lives near Calliope. Next day we said goodbye to the Perretts – they *are* a nice family; they send their regards to you, as they feel they know you, through me – and set off, sliding right off the road every few yards. At Taroom we couldn't get across Palm Creek, which was flooded, so had to be pulled over by a gravel truck. Bill gave the driver two out of our precious horde of pineapples left over from the barbecue, and we went on to Moura, where the big coalmine is. Deciding to deviate and visit the huge dragline, which is the size of a ship, we ran straight into a thick, smelly bog and came to a definite stop. 'Let's have lunch', said Bill there and then, which we did. Eventually, after pushing and shoving ineffectually at the car in the noonday heat, we gave up and walked 2 miles to find a kindly Italian who pulled us out with his truck. So bang went the last couple of pineapples!

We reached Bill's home at sundown with no further mishap. He has invited me to stay a couple of days and see the place, so that is where I am just now. The family originally came from way down south in Victoria – extremely gracious and pleasant people – and the property is rather fine, with lots of lush green river flats, and steep hills covered with trees, where the Brafords – Hereford breeders with Brahman bulls – graze contentedly. They are attractive cattle with glossy red and white coats and floppy briskets.

I slept like a log that night and woke up to a very humid day. Bill has three schoolfriends from Victoria staying here too. They really thought that Queensland was made up entirely of sand-dunes and gibber plains, so Bill and I set out to prove them wrong. We all piled into the tilly and went for a tour, over hills and across creeks full of water, with thick vegetation on the banks. In due course we became bogged (Bill is good at that) and had to abandon the tilly and walk 5 miles home. After a sumptuous lunch we went for a ride and honestly I have never seen a funnier sight than the three Victorians trying to ride stockhorses. One got on by climbing up on the bonnet of his car and jumping awkwardly at the saddle; luckily the horse stayed still, though it stared with white-ringed incredulity at its rider's extraordinary behaviour. After a ride fraught with alternating hysterics and near-misses we turned for home, and their horses took off at full gallop. Bill and I met up with them ten minutes later, all hot and breathless at the yard gates. Why none of them came a cropper is beyond me.

Yesterday afternoon we went out for another ride, and after a while Bill let me have a go on his extra special camp-horse/polocrosse horse/favourite all-round horse: a big, dark brown gelding called Winston who, when he's fed, can easily gallop 8 miles without raising a sweat. I was doubtful if I could manage him, but he was kindness itself to me and a delight to sit on.

Now that I have just met these super people I find myself leaving again. Bill is taking the boys and me up to Rockhampton after lunch, where I hope to meet Sharon. The train leaves at 5.30 and gets into Alpha at some godforsaken hour. It's just as well I got a lift with Bill Stevenson, because there have been bad floods on the coast and no trains are running north or south.

Wendouree, Alpha revisited

18 January

Well, here I am back at Merv's place! It's funny to be here again, though nothing has changed at all. Merv is just the same, his mother is as crotchety as ever, poor old girl, and Drew has not improved much, though he is growing into a handsome little boy.

That nice Bill Stevenson and his friends took me the 90 miles to Rockhampton, bought me a milkshake and a big bag of fruit, saw me onto the train with ten minutes to spare, and went off to a dance. Sharon was already on the Midlander, getting worried that I wouldn't turn up, and away we went on the 272 miles to Alpha, pulling in at 4.35 a.m. after a deadly slow trip, with the blasted air-conditioning making everyone shiver. Neither of us slept much, due mainly to an old drunk opposite who kept saying 'Git up now, Blondie!' to me until, to our relief, he got out at Emerald. Merv was at Alpha station to meet us, complete with new Mercedes, and the sun was up when we arrived here. Jan, the girl who took my place, greeted us with breakfast; she too fell for Doug when he was fencing here earlier, but he is an elusive young bloke with a stubborn independence, so she didn't have much luck apparently.

Later we met Dan, who is working here instead of old Joe, who left a while back. He has a wife Joan, son Frank (14) and 8-year-old twins. They're all mad about horses and they treat each muster as a picnic, which can't be a bad thing. Twinkle is *still* over with the stallion and I'm very fed up, as I was hoping to introduce her to Sharon.

Merv took us out after breakfast to see the new yards Doug and Don built in Charlemont Paddock. There we met Dan with the horses and rode round looking for a few stray head of steers. I rode Sugar, the 3-year-old part quarter-horse filly, who is very quiet and obliging but nervy, shying constantly at her own shadow. At Clara's Dam it started to pour, so upon returning to the Landrover Merv took the twins and Joan home while the rest of us rode the 8 miles, each leading an extra horse. It was the heaviest rain I have ever been out in, with thunder and vivid pink lightning, quite frightening really. Water poured off us and swirled round the horses' fetlocks, and when we came to a tree down over the track and had to leave the road to go around it, down went all the horses, up to their flanks in soft sandy mud. Sugar went mad and floundered all over the place in great panic, which is just a little awkward when one is leading a second horse and trying to hold a saddlecloth over one's shoulders. However, we got home in one piece and after supper slept soundly like fallen trees ourselves.

On Sunday we walked down to the creek and found it way up, about six feet over the road. Dan and Merv extricated me, struggling, from the Landrover and dumped me in the brown water, just for the hell of it, but it wasn't cold. It rained all that day and all the next. We got the camping gear together, but now we aren't going until at least next Sunday because of the wet, so I have written to Glen's sister on the sheep station putting off my visit, as there just isn't time.

On Tuesday Dan, whose car was parked on the far side of the creek, managed to get to town. He made some enquiries about Anakie, but was advised not to go near the place for at least three weeks as it's just one big bog of black soil. Sounds like fun! Merv hopes to be able to take us and the gear down in the 'rover, help us set up camp and return on his own. We can rail the camping stuff back to him plus his .22 rifle, as there are often some dubious characters at these places. He thinks we ought to buy an old banger to escape in if we sense trouble, but I reckon we'd get it bogged and have to run for it anyway.

Next day we rode round Greentrees, my favourite paddock, and still we couldn't leave the vehicle tracks without the horses becoming bogged, though the two mobs of brumbies that appeared seemed to be managing all right. Today we went to the Strip to look at the steers, which are nice and fat. Lunch was eaten at Sandy Creek, where we all went for a swim, fully clothed except for boots; my shirt ended up being ripped from hem to collar at the back, much to everyone's amusement. The twins were tickled pink and could hardly stay aboard their ponies for laughing.

Fossick Creek Camp, Sapphire, via Anakie

26 January, Australia Day

Here we both are, camped on the edge of an almost dry creek and not very much richer than before. Between us we have found well over a thousand small chips of sapphire and zircon, and after some proper digging with picks I unearthed the one and only sapphire of cuttable quality, about half an inch in diameter, blue and green with a few flaws.

We left Alpha on Saturday at 9 a.m., picked up a huge box of tucker that set us back $13, and set out along the Capricorn Highway – a grand name for an incredible road that even goats would find trouble in traversing. Merv's new Landrover bumped and crashed over washouts, rocks, gullies and vast holes. We paused for tinned salmon sandwiches and tea from our gleaming new billy and arrived at Anakie about 2.30 p.m. Six miles out from there we hit the Big Smoke of Sapphire, a 'town' of about four dilapidated houses, one store (with a telephone and fridge) and lots of camps spread out over a big area, almost like the picnic races but not so cordial. Merv asked some people called Gleeson where the best place to camp was, and they said under the two big acacia trees – so here we are now, under a large tarpaulin.

It is quite lovely and not at all as I had imagined a gem-field camp to be: all dust and flies and searing heat. Instead there are the usual eucalypts everywhere, lots of green growth underfoot and a pretty creek just below our camp. Merv helped put the fly sheet up between the two acacias, erected our stretchers and got a load of firewood, then he left for home with his new girlfriend whom he thinks he might marry, with luck. I am doing all the cooking as Sharon hates it; we had steak, onions, pumpkin, carrots and spuds for supper the first evening, followed by fresh milk courtesy of Peggy. Under the next tree is an ex-army pommie with a gigantic black moustache; he's living in a converted Landrover, just touring around as the fancy takes him. He's been very kind to us, taking us out and about a bit, so we are inviting him to supper tomorrow evening in return.

The first night we settled down at dusk, having no lights to read by – and what a night it turned out to be! I lay staring at the massive tree trunk six inches away from me, and it seemed to be moving. On striking a match I saw that the surface was heaving with large hairy caterpillars, all galloping up the stem like vertical lemmings. Surprised but not unduly worried I lay down, and was just going to sleep when Sharon gave a shriek and threw out a grub from her pyjamas. Simultaneously I found two on my nightie and one on my neck; in fact we felt them walking over us all night, until we became used to their tickly feet and began to ignore them, eventually drifting off to sleep.

Some time later I woke, froze instinctively and slowly reached for a match. Sure enough, there was a green and brown spotted snake with a blunt tail disappearing under my bed – which suddenly felt very near the ground. I leapt out, grabbed the shovel and an axe while Sharon fumbled for her torch, and cut off the creature's head as it emerged from under the tent. We found out later it was only a fairly harmless water-snake on its way back to the creek – but isn't it funny how snakes almost always produce an overreaction like that?

The caterpillars were still marching next morning. Presumably they were going to wrap themselves up in cocoons in the branches, but we had an all-out attack on the ones still in the tent, and the meat-ants helpfully carted away the corpses.

We rise at 5.30 every day, just before the sun, and get settled down on the old mullock heap in the creek by 6.30 a.m. Bernie Gleeson lent us a sieve and showed us what to do with it, and we found 57 chips of sapphire that first day, much to our excitement, although they are virtually useless as gems, being too small. Next day we found 208 chips, the day after 279, then 62 (a poor day), then 234, by which time we were both a bit fed up with sitting in the creek, which seems to yield only leftovers from other people's sieves. We swim in the waterhole where fat steers and horses drink, and there is a bore and tank near by for our camp supplies. Merv kindly gave us a round and two big bits of corned beef, which we hang up every night in a tree and wrap up in our swags during the day. It's keeping very well.

John, our English neighbour, took us on a tour of Reward Creek, Rubyvale and the Scrub Lead diggings, but there was nothing on the surface. One is supposed to come prospecting after heavy rain, when the sapphires shine out from the ground, but we haven't seen any, and there was enough rain last week to wash off any amount of dirt. However, we did find a big piece of black jasper in the creek, which Bernie says will make a nice stone. That night we left a bit of meat sitting in a billy strung up to the fire ready for breakfast, but some mongrel dog took off with it in the night so we had to have sardines and poached eggs instead – yuk!

Today John took us up the hill to meet an old Scot called Bill Taylor. He was born in Turriff, came out here in 1923 and has a wonderful, mixed accent. He lives in a caravan and has a hut in which he cuts and polishes agates with his own machinery. He gave me two bits of topaz, fed us with tea and biscuits, corned meat sandwiches and fresh peaches from his little fridge, and was so pleased to have someone to talk to and entertain that he wants us to go and see his movies tomorrow night.

Farewells: last letter from 314 Queen Street

2 February

Last Friday night a strong wind blew up and lightning flashed. Sharon and I both woke at the same time and shot out of bed, Sharon to retrieve the corned beef and our breakfast hanging in the tree, and I to grab my bed and stumble over the pile of firewood in the middle of the tent – all in the dark – to move it under better shelter. We got back into bed panting – and not a drop of rain fell. We did feel foolish.

Saturday was an easy day spent walking 2 miles up the creek to find a beautiful sandy waterhole, where we swam and lazed about. In the evening John the Warrant Officer came for supper. We gave him boiled corned beef and pumpkin (but I did some carrots too in case he didn't like pumpkin, which he didn't), spuds in their jackets and a mess of saffron-type rice mixed with a tin of condensed vegetable soup and fried onions. You try making fried rice in a tea billy – it's not easy. He loved it all and was most appreciative. After dark he took us up to Bill Taylor's caravan, where we were shown film of Roma, Dirranbandi and various other places. Much later that night I woke at about two to the sounds of a very lively party over the creek, which seemed to break out into a full-scale brawl. I went back to sleep with one hand on Merv's rifle.

On Sunday morning two bearded weirdies (English hippies)

arrived and camped right beside us. They were wearing sarongs and elephant-hair bracelets – not a pretty sight – but they left next day, thank goodness. Then Merv and Betty turned up on their way back from Springsure, much to our surprise, very kindly bringing with them a sirloin, minus the fillet, and some fresh eggs. We fell upon the meat for supper and gorged ourselves like wild pigs, it was so good. Bernie Gleeson presented us with fresh mangoes, which finished off the meal to perfection.

On Tuesday John the Pom took us and old Bill Taylor out to Black-water, about eighty miles away, to look for some petrified wood and ended up having to buy some as there was none to be found. It was a boiling hot day and we had a highly indigestible lunch of melting sardine sandwiches and black billy tea in scant brigalow shade before going 'home'.

After a cheese and onion omelette, of all things, on Wednesday morning, we broke camp, dislodging masses of caterpillar chrysalises, which brought the meat-ants running in hundreds. We tied up all the gear in the tarpaulin, then spent the rest of the day with Bill Taylor, who showed us how to polish various agates and then presented each of us with a completely hand-made pendant of polished pink and grey agate on a silver chain; he really is a dear. Another old prospector turned up and gave us a few grains of real gold that he had found one day, together with a handful of zircons. At five we went along to Bernie and Nola Gleeson's house to have a shower in the most wonderful shower-house you ever saw: three corrugated-iron sheets made up three walls and the fourth was open to all, though it did face the house and not the open countryside. A bucket with holes in its bottom and a wire for control of flow was strung up above, then one grabbed a towel and ran round three sides of the house to get in, through burrs and prickles and dust. We were invited for a meal before catching the train: delicious mince with capsicums and rice, and custard pie. The Gleesons were a nice couple and I'd have liked to get to know them better, but once again time forbids much socialising. John the Pom took us to the station and the train left at 8.30: a real old-fashioned mail train with comfortable leather seats for us to stretch out on, as there was no one else on board, or so it seemed.

Fourteen hours later we hit Rockhampton, had a slap-up meal and went our separate ways: Sharon north to Atherton, while I set off for Gladstone on the Sunlander, having bought myself Dean Martin's record of 'L'il old Wine Drinker Me' – nothing significant in the title, I just like the song. At Gladstone there is a vast alumina plant employing people of every nationality, all rather terrifying to a mere bushie. I left the Sunlander there and caught the mail train to Maryborough, after trying unsuccessfully to ring Jean.

River Road, Tinana (Ruth's home)

9 February

On Monday I had my first cholera jab. On Tuesday morning the budgie woke me by banging furiously on his bell until his cover was removed, by which time my arm was quite sore and inflamed. However, I managed to go to the dancing class that night, where I was accused of using a blonde rinse on my hair; in fact the streaks are the natural result of not wearing a hat for the past year.

Ruth Andersen and I took Kirstine Perkins out to dinner at the Caravilla Motel last night; Kirstine is one of the two dancing teachers at the class, and a prettier, more delightful girl I have never met. We drank a bottle of Barossa Pearl between us and giggled helplessly – it seems to have that effect on me. Later we went to see the film *Khartoum*, starring Charlton Heston, but it was *so* long, drawn out and boring – talk, talk, talk, a brief charge of camels and horses then talk, talk, talk again. At school my notes on the incident took up half a page; maybe that helps to explain why I failed so miserably in History!

Just a short letter this week. The next one should come from Sydney where I hope to be in a few days' time.

8 Shannon Street, St Ives, Sydney

16 February

As you see, I am in Sydney now, staying with Peter and Rosemary Rich and their 2-year-old son David, who live very pleasantly in their charming white suburban house surrounded by lots of green trees and grass. But I am very homesick for Queensland. I miss it so much, and all its *nice* people and fine animals, that I seem to have definite withdrawal symptoms.

A salesman friend of Jean's gave me a lift down to Brisbane after Jean and Ruth saw me off at noon, both on the verge of tears. Mrs Sinclair invited me to stay the night with her when I rang her up earlier, and the salesman drove me right up to her house in Wavell Heights at six that evening. It was lovely to see her again, and funny to think that it is almost two years ago that I first came to this great country. Next day I 'finalised me bankin', getting a good collection of travellers' cheques to spend in Hong Kong, said a sincere goodbye to Mrs Sinclair – a truly fine person – and went to meet Kirstine Perkins, the dancing teacher, for a fabulous Chinese meal. Afterwards we went to see Rex Harrison in that strange film *Dr Doolittle*, then she came with me on the tram to the station to see me onto the Sydney Limited Express with five minutes to spare. Goodbyes again, and I settled into my *sleeper*; I thought I'd have a change from the floor. Anyway, one couldn't sleep on the floor in New South Wales , it's far too civilised a state. The cabin was comfortable, complete with shower, loo and drinking water, and a blue-rinsed woman got on at Casino to occupy the lower bunk.

I woke at sun-up and had a sheet of rump steak for breakfast in the dining bar, then we chuffed into Sydney Central at 8.30 a.m. What a seething mass of humanity Sydney is. People eyed me warily when they saw what I was carrying – Ian Swinbourne's speargun. From the station I rang Peter and he told me to hop onto another train and head towards St Ives and he would meet me. It's ten years since I last saw him in Ceylon, and then I was only ten myself, so obviously we saw a great difference in each other. It's good to see him so happy with Rosemary. He seems to have everything he wants around him.

Oh, I haven't told you about the terrific weekend I had before leaving Maryborough. My last day was spent *mustering*, a real bonus as I had firmly believed I'd hung up my spurs on leaving the bush. It came about because on Saturday night Ruth and I dressed up and went to town looking for a dance. There was nothing doing and we were on the point of going home when we discovered there was a ball on 27 miles out. Off we went, and the first person we saw swinging through a gypsy tap was Toby the small ringer – minus his hat this time – the tall Texan and their mate Barry Shaw. Barry asked me up for a jive, which we did most efficiently, and at one point said he was going mustering the next day. I said 'Can I come?', just in fun, and he looked at me as if I was nuts and said 'Of course'. He couldn't believe that I really *liked* riding and herding cattle in rough bush country, as all the town girls he knows tell him to get lost when he mentions mustering.

The next day of course it poured with rain, but Barry and Toby came round in an ancient Austin truck (with holes in the cab roof for wartime guns) to collect me and a load of firewood from the mill at the same time. Then we drove out to Howard, 18 miles north of Maryborough, where they live. Barry's place is half a mile out from the little town, with a few acres of healthy citrus trees, and his father breaks in horses for people as a sideline. We had an early lunch while Barry's younger brother got the horses in, and I was given an enormous racehorse with legs a mile long. I must have shown my surprise because Barry said anxiously, 'You'll be able to handle him won't you?'. As it turned out, the only fault the lovely animal had was that he shied sideways quite frequently, but otherwise we had a super ride, with lots of cantering through dripping trees after cattle. The rain never let up, but it was still a lovely day. After a late supper with the family Barry drove me home in his Zephyr and confided that he wished he'd met me a long time ago. He's only 19, but a great kid.

I have to have one more injection, collect my tickets, then it's just a matter of waiting for the ship to sail. See you soon!

28 February

Only two and a half days left in Australia. Actually I might as well be in England, for since leaving Queensland nothing is the same. Sydney is a stuffy, artificial place and Canberra, which we visited last week, is even worse, though more beautiful and peaceful with its lack of industrial noise. The capital city seems to be made up of large Embassy houses, vast, unfenced gardens and of course Government House and the man-made waterhole in the centre, Lake Burley Griffin. It was really hot there – 102°F in the shade – while we were staying with Rosemary's parents, who incidentally live next-door to the Ceylonese Ambassador. The country is in the grip of a pretty severe drought, with water restrictions. Peter and Ro took me to Bateman's Bay, on the coast 100 miles away, through much dead dry country, with the cattle and sheep looking crook from lack of food and water. We had fish and chips on the beach, but the water was freezing so none of us ventured in for long.

It was an amusing dinner that night with Peter's in-laws. I'm afraid I put my foot in it when someone asked what kind of wine they drank up north and I replied 'Barossa Pearl', nostalgically. Everyone looked down their noses as if I'd said something truly shocking.

The next day Peter took me round the War Memorial and museum, which is well laid out, with big oil paintings, remnants of guns and submarines, and five aeroplanes downstairs, including one huge, black Lancaster bomber in all its sinister glory.

On Sunday one of Ro's sisters took us for a sail on a yacht. I hopped overboard into 60 feet of water and had a good swim, but then a gust of wind came along and away went the boat. It came back just as a floating mass of weed attacked me, and I shot back on board in a state of near panic. That night I met the Ceylonese Embassy people, a Mr Couret and his Scots-born wife, whose houseboy has a brother working on my godfather's tea estate where I am going to stay next month. Once again, it's a small world.

On Monday we went for a swim in a weir among thick pine trees, returning to Sydney next day in time to see, on television, Lionel Rose, the Aboriginal bantamweight, win his fight in Japan. I also bought a copy of *Hoofs and Horns* and read about Doug Spann being all-round champ at one rodeo in Victoria. I wonder how he is.

Now it is Friday. This afternoon we take my ports down to the wharf, and as the *Oronsay* gets in at noon today we should see her tied up near the bridge.

SS Oronsay, *passing* Cape York

7 March

Here we are, three days out of Sydney and still on the Queensland coast. The ship passed Cairns at two this morning and I felt very tempted to abandon ship and swim ashore. Part of me will always remain in that town.

The old ship seems to have shrunk since I was last on her, or perhaps I have grown somewhat since the age of five. The passengers are, to use a Queensland expression, a scungy mob, except for one English girl, Christine Ashworth, in the same cabin as me who has been working in New Zealand for a couple of years. Last night I was adopted by a bearded Yank who smokes marijuana and he now wants me to play ping-pong with him – what a thrill.

Peter came with me by train to the docks, then we walked down George Street to the ship: a most original and peaceful way of arriving, I thought, instead of the usual racing taxis and luggage everywhere. As my three cases had gone aboard the Friday before, all I had to do was show my ticket and climb up the gangplank. I didn't see Ian or Murray Swinbourne at all. Ian rang up in a state, saying he was on duty all weekend, which was a shame, but he kept me entertained on the phone for about half an hour and arranged to call for his speargun some time soon.

You'll never guess who turned up on Saturday – Dick Perrett, from Bottle Tree Hills, Wandoan. I nearly died on answering the doorbell to see him standing there with his trousers at half-mast and his bush hat pulled down over his eyes. He had flown south that day after mustering until noon, especially to see me off and also to visit some relations. He took me to the Taronga Park Zoo that afternoon, after which we boarded a ferry and crossed the harbour, chatting non-stop about Queensland and how much nicer it is than any of the other states in Australia.

He came on board to see me off, clutching a packet of scorched almonds as a parting gift, conversing in his usual bellowing voice about his breeders having abortions, the cursed dingoes and all the other ordinary bush subjects. I couldn't help smiling at the outraged expressions and raised eyebrows of the people around us, but it made me sad too. Peter had to go back to work at 10 a.m. but Dick stayed on until 11.30, when streamers were thrown from ship to shore, the whistle blasted everyone off their feet and the old ship pulled out, slow astern. I must admit to shedding a few tears and being unable to face lunch, which was served almost immediately. We ran into a storm out from the Heads, but the stabilisers were so efficient we rocked around very little.

The passengers are mostly Kiwis and South Africans and drab Liverpudlians. However, at my table there are two Roman Catholic priests going home on leave, both based at Rockhampton. One of them, who's Irish and has a terrific sense of humour, was out at Alpha for a couple of years – a small world once again. Beside me is a South African lad who jackarooed near Blackall and breeds racehorses back in South Australia, so it is quite an interesting table, complete with a broad-accented Scots couple from Dundee, a young Chinese pair on honeymoon and two Chinese nurses going to try their luck in London.

We are just about at the tip of Cape York, having gone through the Reef yesterday, when I recognised South Molle and all the places the *Sue* visited on that fabulous trip two Christmases ago.

Must close now, as Chris is nagging me to go sunbathing. There is a water restriction on board at present because of the long haul to Manila.

Friday Last night Chris and I were invited to an officer's cocktail party in a dimly lit cabin. There were canapes and things, and lots to drink, and about ten officers, some of whom were okay but the rest obviously doing their duty and not enjoying it much. After that there was a 'Reef Night', dancing on the top deck until the early hours and swimming and sitting in deckchairs to recuperate at intervals. At 1 a.m. when the music stopped, Chris and I and some boys went for a swim; it was a lovely balmy tropical night with a warm wind blowing. Unfortunately one of the Kiwis took exception to a

Scots pastry-cook's long hair while in the pool, so they got out and started fighting. The pastry-cook nearly went over the rail at one point, but the Kiwi was coming off rather badly when the Master-at-Arms came and stopped the fight. Kiwis are worse than Aussies for fighting apparently, and that's saying something.

Sunday   Last night was the Captain's cocktail party, which was quite fun except that I was introduced as Barbara Hardwick, for some extraordinary reason best known to the Purser. At dinner the place was decorated with streamers and paper hats and a gala dance followed, during which I made an abortive attempt to dance the cha-cha with a crocodile shooter from the Northern Territory.

The food on board is very good: far too much of course, and I'm putting on the half stone I lost in Queensland. The unfortunate priest who sits next to me at table, Father O'Connell, has received three letters from various nasty women asking why it is that a priest should be drinking alcohol, and declaring it's disgraceful that a man of God should succumb to the evils of liquor. Poor bloke! What hypocritical, unchristian ratbags some people are!

I must stop and post this epistle, as letters have to be in by tonight for posting in Manila on Tuesday.

P.S.   A telegram was waiting for me on my bunk when I came aboard at Sydney, which said, 'Good sailing. Return some day. Hassall, Rankine'. I'll bet they sent it over the transceiver and now all the neighbouring stations will know that Spear-grass Sue the jillaroo is on her way back to Scotland.

DEDICATION
This book is dedicated to the fond
memory of Glen Rankine, my
immediate boss and good friend on
Meadowbank Station, who was killed
in early 1969 when the ill-fated
chestnut pony Imp fell on her while
out mustering.

Glossary

bail framework for securing a cow's head during milking.

bangtail (vb) to remove the long tuft at the end of a beast's tail so that any cattle who escape the muster can be recognised from a distance and brought in next time.

blackleg an infectious, generally fatal disease of cattle and sheep that is picked up from the soil and characterised by painful, gaseous swellings in the muscles.

bosal American bitless bridle.

buckjump rodeo event in two sections, saddle bronc and bareback, where a bucking horse has to be ridden for 10 seconds and 8 seconds respectively.

bulldogging rodeo event, called steer-wrestling in America. Rider tries to throw a galloping steer by leaping off a horse, which runs alongside, and immobilises the beast by tying a front leg to a back leg with rope.

bull ride rodeo event. Bucking bulls have to be ridden for 8 seconds.

bushed lost.

buster fall from a horse.

camp-draft (vb) to cut out, on horseback, individual beasts from the 'camp' or held mob. Camp-drafting is also a national bush sport where points are won or lost depending on the horse's (and rider's) performance round a set course in a certain time.

cattle spread cattle station (US).

chook chicken, hen or rooster.

cleanskin a beast with no brand or earmark.

cradle metal framework in which animals are held firmly to be shorn or given routine treatment such as branding and marking calves.

crush   a narrow, fenced, funnel-shaped passage along which cattle are driven for handling.

dogger   person who hunts dingoes for a living, claiming the bounty on their skins.

Droughtmaster   beef cattle bred from British shorthorn/Devon red and tick-resistant Brahman, to withstand drought and tropical conditions.

duffing   the stealing of stock.

feature horse/bull   one particularly hard to ride. Competitors' names are put in a hat and the first drawn out earns the chance to win big prize money if he or she can stay with the horse or bull for the allotted time.

feral   term describing domestic animals grown wild and living in the bush, e.g. feral pigs, cats; a brumby is a feral horse.

galahs   pink and grey parrots. 'Galah session' refers to a certain time allotted to far-flung ladies to talk to each other over the transceiver.

gilgai   shallow waterhole after rain.

Hereford   very productive, hardy, early maturing breed of beef cattle, originally from Britain, with red body, white face and other white markings.

Hogmanay   New Year's Eve (Scottish).

jackaroo/jillaroo   young person who lives with the family and learns how to run a cattle or sheep property from the bottom up to managerial level.

Kelpie   Australian sheep dog.

killer   beast shot and butchered for meat.

Lippizaner   fine breed of white horse used especially in displays of dressage.

mickey   young cleanskin bull that missed being castrated.

night-horse   horse kept near the homestead overnight in order to bring in the other horses in the morning.

pigroot, root (vb)   (of horses) to buck or kick up the back legs.

poddy   orphan calf, foal or lamb, hand-reared.

points   rainfall was measured in pre-metric days in points, there being 100 points to an inch.

poley   Australian stock saddle with thigh and kneepads.

preg-testing   testing for pregnancy

prop (vb)   (of horses) to stop suddenly with all four legs stiff, jolting the rider.

quarter horse   horse bred in the US for speed over short distances, originally a quarter of a mile, and used for roping and cutting out cattle.

rickety   staggering.

ring-barking   method of killing trees as they stand, as opposed to pushing them over with a bulldozer during land-clearing operations.

ringer   station hand, especially a stockman or drover.

roughrider   one who breaks horses to the saddle; one accustomed to rough riding and who competes in Australian rodeos.

round   standard cut of beef from the hindquarter.

running a banker   creek or river flooded to top of bank level; about to overflow.

scours   persistent diarrhoea in animals.

scrubbers   wily cattle who live in the deepest scrub and manage to avoid being mustered for years, if at all.

shade tree   tree planted, retained or valued for the shade it gives.

shorthorn   breed of dairy or beef cattle with white, red or roan markings, having short horns.

smoko   morning or afternoon tea break.

spavined   of horses with disease of the hock joint in which there is a hard bony tumour or excrescence.

spayed cow   one that has had her ovaries removed.

stock route   a designated route, usually 1 mile wide, along which stock are allowed to be driven.

strangers   cattle belonging to neighbouring properties that have strayed through the boundary fence.

surcingle   band around horse's body, especially to keep blanket, pack, etc. in place.

tilly   utility truck , usually a car with two or three front seats and an open tray back.

toey   refers to restless, fearful cattle 'on their toes' and ready to charge.

twitch   'bushman's chloroform', a thick stick or axe handle with a loop of rope at the end that is attached to the horse's upper lip and twisted tightly to prevent movement while an operation is in progress.

vealer   eight- to nine-month-old calf in fat condition.

white ants   termites.